HEROIC TALES *of* WETLAND RESTORATION

D0756805

HEROIC TALES *of* WETLAND RESTORATION

Esther Lev

Produced with generous support from:

Oregon Department of Fish and Wildlife

Oregon Division of State Lands

Pacific Coast Joint Venture

Szekely Family Foundation

US Environmental Protection Agency
Region 10

US Fish and Wildlife Service

Published by The Wetlands Conservancy
Tualatin, Oregon

Library of Congress Control Number: 2001096007

ISBN 0-9660405-1-1

Printed in the United States of America

CONTENTS

ACKNOWLEDGEMENTS

Traveling around Oregon, I learned many new things and discovered new ways of looking at familiar landscapes. My travels showed me that maintaining water, topsoil and vegetation is the landowner's best insurance policy. It allows them to continue living where they are and loving what they do. I learned how land management evolves over time and responds to new understandings of the relationships between people and their natural environment.

I learned about cattle and sheep ranching, dairy farming and grass seed production. I began to see opportunities for integration between wetlands and working landscapes, people and wildlife.

I found that private landowners and government agencies follow a more parallel path in recognizing environmental problems than either might think. Solutions can become the departure point, fueled by the fear of change and the cultural differences between the bureaucracy and the ranch. Yet people in all parts of the state crossed hurdles to forge the magnificent solutions contained in the tales, photographs and paintings in this book. Each success pays tribute to the power of partnership between individuals, agencies, land, water and wildlife. Each hero can inspire us all.

First and foremost, I thank the heroes in this book for their noble work, their modern tales of change. Their warmth, honesty and creativity inspires me. I hope other land-owners find similar inspiration to restore and enhance their wetlands.

I greatly appreciate the trust of the following project donors, whose support enabled the interviews, writing, photography, art, design and publication of this book:

Oregon Department of Fish and Wildlife

Oregon Division of State Lands

Pacific Coast Joint Venture

Szekely Family Foundation

US Environmental Protection Agency
    Region 10

US Fish and Wildlife Service

As I searched for examples of wetland restoration around the state, many people shared their knowledge of people, projects and places. Without their help I may not have found these wonderful stories. In particular, I thank Rob Tracey, Natural Resources Conservation Service; Ann Donnelly, South Coast Land Conservancy; Faye Weekely, Curt Mollis and Jim Hainline, US Fish and Wildlife Service; Bruce Taylor and Sandra Fife, Oregon Wetlands Joint Venture, and Bianca Streif, US Fish and Wildlife Service.

In addition to the coaching and editing provided by Dawn Robbins, I thank Janet Morlan, Bianca Streif, Bruce Taylor, Anne Donnelley and Thalia Zepatos for reading and commenting on the manuscript.

The photographs in this book make the stories come alive. I want to thank Madeleine Blake, Richard Wilhelm, Steve Roundy and Deb Stoner for these outstanding photographs. Thank you to Curt Mullis and Teresa DeLorenzo for providing photos of several of the projects. Steve Katagiri's marvelous illustrations remind us of the plants and animals that benefit from these wetland restorations. Thank you to Laurie Causgrove for bringing together all the pieces and making this beautiful book.

INTRODUCTION

Wetlands are vital to our lives. They store, clean and filter our water, prevent soil erosion, and control flooding. They provide rich habitat for thousands of species of birds, fish and mammals. And they are indescribably beautiful.

Yet for more than two centuries, we have destroyed millions of acres of wetlands for agriculture, commerce and other developments. In today's world, we hear tales of people in rural areas abusing the land. We see photos of landscapes with denuded riparian zones, deep-cut, incised channels and eroded stream banks. By the same token, rural residents tell us how government agencies take away property and bureaucrats care more for animals and plants than for people.

Heroic Tales of Wetland Restoration sheds new light on these stories. A dozen rural landowners tell about how they have changed farming methods to reclaim wetlands, streams and rivers. Some of these pioneers have bumped heads with government, but most have also forged partnerships with hard-working agency staff who have helped them maintain a rural life and breathe new life into their precious land.

The real-life heroes include both farmers and public employees. Together, they have restored natural Oregon landscapes from the Columbia River to Cape Blanco, and from Bonanza to Bear Valley. They have rejuvenated oxbows, lush with sedges and cattails. Sandhill cranes, black-necked stilts, blue-winged teal, cutthroat trout and Nelson's checkermallows find homes in newly restored habitat. Many fish, aquatic animals, birds and plants thrive in places where they once would have withered.

The tales in this book are not fairy tales. They are true stories about passionate people who have overcome obstacles. Their hurdles include red tape, resistant neighbors and physical challenges. Jerry Hines, of Chiloquin, complains about the lack of a Prince Charming to steer projects through an unwieldy course that seems interminable. After seven years of multi-agency confusion, Doug McDaniel says he may give up on his Wallowa River project. Others, like Bonanza's Louis Randall, found the special helpers they needed.

Why did these heroes continue down the bumpy path to restoration? Edith Leslie, of Beaver Hill, wanted to transfer to her children the land that has been in her family for more than a century. Mark Tipperman and Lorna Williams fled from the rat race in Snohomish County, Washington, to create a peaceful life for themselves in the Blue Mountains. Mark Knaupp, of Rickreal, longed for ducks on his property.

In the end, each hero can feel good about creating more Oregon wetlands. "We are caretakers of this earth for such a short time," says Sharon Sinko of Myrtle Point. "We would like to give something back to land that has been good to us."

Their tales are part of a larger national story. More than 75 percent of wetlands in the lower 48 states are privately owned, making landowner stewardship a critical part of a wetland conservation strategy. In 1985, the Natural Resources Conservation Service created several landowner incentive programs. Two years later, the US Fish and Wildlife Service initiated its Partners for Fish and Wildlife Program.

As these programs change, so do the people whose lives they touch. Residents of Oregon's rural and urban communities are gaining a new understanding of how wetlands enrich our lives and protect our future. Landowners and members of public agencies, non-profits and citizen groups preserve and conserve quality wetlands and restore others.

In the first section of this book, readers can enjoy and learn from these tales of vision, passion, perseverance and economic survival. The second section of this book describes land conservation options and a range of state and federal technical assistance and funding programs. It also lists and explains the programs the people in this book used and some of the regulations that governed their work.

The last section in the book includes descriptions of the restoration techniques employed by the landowners, as well as recommendations for the future. It outlines difficulties experienced by landowners working with federal and state incentive programs and includes landowner recommendations for ways the programs can better accommodate their needs.

MAP OF PROJECTS

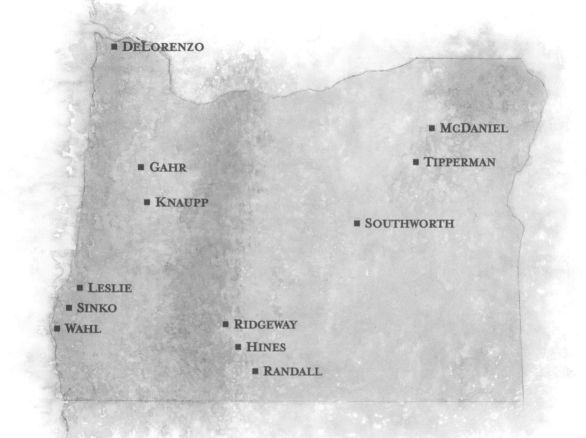

TED GAHR

We need the tonic of wilderness, to wade sometimes in marshes where the bittern and

meadow–hen lurk, and hear the booming of the snipe; to smell the whispering sedge

where only some wilder, and more solitary fowl builds her nest and the mink

crawls with its belly close to the ground.

HENRY DAVID THOREAU

1817 – 1862

LAND CHANGES
VACATIONERS' LIVES

Ex-Californians pursue wetland education

McMinnville, Oregon

Ted and Harriet Gahr had no intention of buying a farm on their 1966 vacation to Oregon. However, the piece of land along the Yamhill River caught their eyes. In 1967, they pulled up their roots in California, bought the farm, and moved to the Willamette Valley. Back then, the young couple had never heard of red-legged frogs, Nelson's checkermallow, or Fender's blue butterfly. Now, the Gahrs share their 350-acre conservation farm and forest with these native creatures and plants.

"It was the scenery, rather than the farming potential, that sold us on the place," Ted said. "Initially, we were more than happy to accept guidance and farming tips from the seller." After farming for a number of years, they leased the land to another farmer, but he gave up his lease in 1992. "Our lessee, even with the help of the subsidy, was having a hard time making a profit on our Muddy Valley soils,"

Ted recalls. So he began looking for alternatives for the land beside conventional farming.

"About that same time, I spotted 300 ducks in pools created by the river overflowing into the cropland. This sparked an interest in restoring the wetlands for the ducks, which led to substantial management changes which are still in progress," Ted says. "Learning over time, I developed an appreciation for many other species of native plants and wildlife that inhabit the farm."

Preserving and enhancing the natural habitat potential of the property became a priority in management decisions that followed. The farm and forest activities now focus on sustaining and improving natural habitat, operating a bed and breakfast business, and opening the property for educational workshops and retreats.

Ted learns whatever he can wherever he can. He has received advice from the Yamhill

"Mostly, I have learned by doing,

experimenting with design,

engineering and construction."

County Natural Resources Conservation Service (NRCS), Ducks Unlimited, and Oregon Department of Fish and Wildlife (ODFW). He reads a lot, visits farms, attends conferences and workshops. "But mostly," he says, "I have learned by doing, experimenting with design, engineering and construction."

Ted signed up with the Agricultural Conservation Program, an ODFW landowner cost-share program to create shallow water habitat for wildlife. He received funds to conduct a land survey and construct a dike and water-control structure. Ted provided the labor. The two-year, 12-acre project was the first of many restoration projects, but the last time Ted received government engineering services.

Over the next three years, Ted continued his wetland and stream enhancements with funds from the USDA Environmental Quality Incentives Program (EQIP). "That's what launched my career in water system and restoration design and engineering," he laughs.

In 1997, Rob Tracey, of NRCS, introduced the Gahrs to the Wetlands Reserve Program (WRP), which would pay them to stop farming and restore the wetland. In January 1999, after two applications to the program, they enrolled 119 acres.

Once the project was accepted into WRP, things moved slowly. "The process took longer than we expected, with glitches along the way," says Ted, "but it was an essential step in continuing the quest for the holistic management of the farm."

Opportunities for habitat restoration are continually unfolding. Along a channeled stream, they found vivid pink clumps of Nelson's checkermallow, a flower listed as "threatened" under the federal Endangered Species Act. The sighting spurred them to create a checkermallow reserve.

Ted proudly points to the 200 checkermallows now growing along the banks of a series of small ponds he built for red-legged frog recovery. He feels commercially propagating the checkermallow through native plant nurseries would have a positive effect on recovery, but has had little encouragement from the regulatory agencies due to controversies about the propagation and sale of endangered species. "Maybe in the future," Ted muses.

Ted believes that with holistic goals and management, forests,

Visitors find peace beside this babbling brook.

good yields of grain without the use of chemicals or fertilizers.

Overall, he feels government programs are heading in the right direction and have a positive influence on improving the health of our watersheds. He also feels continuing refinements are necessary to make programs more effective and efficient.

"The Wetlands Reserve Program made it possible for us to continue pursuing our goals of preserving and enhancing the natural habitat values of our property," Ted says.

fields and wetlands can provide large amounts of food crops for wildlife and people. He continues to experiment with cultivation of a constructed 30-acre wetland, which serves as winter habitat for dabbling ducks and produces

On summer days, red-winged blackbirds alight upon cattails.

TERESA DeLORENZO

An old pond —

a frog tumbles in —

the sound of water.

MATSUO BASHO
1644–1694

COMMUNITY EMBRACES
SLOUGH REVIVAL

Project lures red-legged frogs, salamanders and salmon

Knappa, Oregon

For years, Teresa DeLorenzo advised others on how to restore and manage their land. "One day," she says, "I realized it was time to purchase and manage my own property – walk my own talk." In 1998, she bought 86 acres along Warren Slough, about 15 miles from the mouth of the Columbia River. Neighbors, Teresa says, were initially skeptical of her – a wildlife consultant from Portland. Teresa quietly began restoring her property. Then the project turned into a community effort.

"The project is visible from the road," she explains, "so neighbors would constantly stop by, watch and offer advice."

The Lower Columbia River Estuary Program agreed to fund Teresa's project. After long discussions and several field visits, officials at the US Fish and Wildlife Service's (USFWS) Partners for Fish Wildlife Program followed suit and provided funds to help Teresa create a passage over a dike that enables fish to swim to a large pond and several streams that drain into Warren Slough.

Teresa's wildlife and advocacy background helped turn the tide in talks with USFWS. "Fish and Wildlife has an anti-impoundment stance," explains Teresa. "It was up to me to demonstrate that changing the 25-year-old system would improve habitat for fish and wildlife." Teresa was frustrated with the narrow focus on fish. Her vision was filled with birds, reptiles and amphibians.

Teresa reached out to others as she pursued the project. "I learned to ask everyone for advice, help, and ideas," she says. When she discovered she needed permits, she hired Mark

"I totally underestimated the community interest and extent to which the project would benefit from community involvement."

Barnes, a consulting planner. He prepared permits for the US Army Corps of Engineers and Oregon Division of State Lands, and negotiated design features and engineering details with the county Public Works Department and National Marine Fisheries Service (NMFS). Ducks Unlimited prepared engineering drawings required for the permit applications.

The first step was removal of a collapsing 18-inch culvert. Teresa paid for the materials. Clatsop County Public Works provided the labor. "I ended up with a three-foot culvert rather than the four-foot one I had wanted," she says, "but was charged less money." She was pleased with the county's hard work and flexibility: "It was a great partnership."

Next, she and her consultant worked with national and state fisheries specialists. NMFS helped her determine dimensions of the steps and ponds. "The local Oregon Department of Fish and Wildlife fisheries biologist, Joe

Sheahan," says Teresa, "was helpful in working through design details."

This restoration was not the usual Clatsop County building project. Astoria's Vinson Brothers Construction did "on-the-ground research and development," as they repaired the dike, built the fish passage, removed culverts, smoothed new stream channels, and built the water control box, Teresa says. Vinson Brothers supported the project by charging only what the grant construction budget would allow.

Next, Teresa worked with Keith Fitzgerald of Alder View Natives in Wilsonville to choose plants for the site. She bought 425 shrubs and 60 Sitka spruce trees from the company — far more than she could haul with her small pickup. "I showed up for 7 a.m. coffee at The Logger Restaurant in search of suggestions of where to find a large enough truck," she recalls. Autio Company, a local manufacturer, provided a truck and driver for two round trips from Knappa to Wilsonville — at no charge.

Volunteers from the Nicolai-Wickiup Watershed Council unloaded and planted the shrubs and provided lumber that had been donated by Willamette Industries to build the water control box in the dike.

The non-profit Northwest Ecological Research Institute managed the grant and provided technical assistance. With their help, Teresa created habitat for some of her favorite

Fish will pass through this new channel to a large pond and several streams that drain into Hall Slough.

animals: red-legged frogs, salamanders, purple martins, and others. At the same time, her efforts helped four species of salmon and sea-run cutthroat trout.

Looking over the land, the wildlife consultant reflects upon partnerships that included truckers, bureaucrats, vendors, engineers, biologists, 4-H, and the local community college: "I totally underestimated the community interest and extent to which the project would benefit from community involvement."

Teresa can adjust water levels by removing a board in this water control box.

In the future, hundreds of salmon and trout will use the resting pools being created in the fish passage.

MARK KNAUPP

In the swamp in secluded recesses

a shy and hidden bird is warbling a song.

WALT WHITMAN, 1820

RESTORATION PROVES LUCRATIVE

Farmer-biologist finds "best of both worlds"

Rickreall, Oregon

Polk County farmer Mark Knaupp watches the habitat change in the bottomlands of his expansive grass seed farm. In the past five years, Mark has restored 320 acres of wetlands and created an additional 60-acre wetland mitigation bank. "It's amazing how the vegetation comes back and how the habitat develops," says Mark. "It's happened very fast. The birds found it and moved into it very quickly."

Mark suspects that neighbors laughed when he and his wife Debbie bought the 200-acre damp fescue grass field in 1976. He was the only interested buyer. Now, 25 years later, some who may have snickered would have to view the Knaupps' purchase as a good investment. They now own 1,200 acres — 400 acres of restored wetland and the rest under cultivation.

"I enjoy waterfowl and waterfowl habitat," says Mark, an avid duck hunter with a degree in wildlife from Oregon State University. "That's what got me started on my first restoration project. Now look!" He points to phalaropes poking their long bills into the mud at the edges of the wetland.

In 1992, he embarked on his first restoration project — creation of a 20-acre shallow pond for waterfowl. In 1995, after a series of wet years and gaggles of hungry Canada geese feeding on his grass each winter, Mark's lands along Mud Slough became increasingly difficult to farm. He decided to enroll 320 acres of the bottomland into the Wetlands Reserve Program (WRP). He used the easement payment from the WRP to buy an adjacent 180 acres. This allowed him to acquire more productive farmland, less sensitive to the elements, and essentially trade it for prime wildlife habitat.

Mark harvested his last commercial crop of tall fescue seed from the bottoms in 1996.

The site was flat and required little excavation. To prepare the soil for native plants, he scraped it, planted it with an annual cover crop, flooded it, and sprayed Round-Up in the spring and fall. Then, without tilling, he planted meadow foxtail and the once widespread tufted hairgrass.

In five years, thousands of native plants, including some rare species, have sprung up on their own and turned the wetlands into a tapestry of color and texture. The purple and white popcorn flower and veronica, dark green tufts of sedge, bulky cattails, and delicate tufted hairgrass provide habitat for a variety of wildlife. The showy pink Nelson's

This former grass seed field is now home to diverse wetland plants and wildlife.

The elusive Virginia rail pokes for food in the Knaupps' wetland.

checkermallow is listed as "threatened" under the federal Endangered Species Act.

Mark figures he and his brother spend a combined two weeks per year maintaining the wetland, primarily removing and managing reed canarygrass, purple loosestrife, blackberry, and Canada thistle. Though the process has

required patience, Mark has no regrets. "Buying the land, farming it and then enrolling in the WRP and restoring it to wetland was a great business decision."

Mark also enjoys the priceless beauty that comes from his decision. Every fall through spring, thousands of ducks and geese, shore-birds and swallows return to the wetlands. Mark has seen birds breeding, including black-necked stilt, Wilson's phalarope, and 11 species of waterfowl. In addition, bitterns, rails, herons and egrets feed in the marshes, while bald eagles and northern harriers work the skies overhead. Mark's list of birds includes some that are rarely seen in the Willamette Valley, including yellow-headed blackbirds and white-faced ibis. "My bird list matches, if not surpasses, the nearby Basket Slough Wildlife Refuge," Mark beams.

Mark's 23 years of farming, combined with his education and passion for waterfowl hunting, gave him the background and motivation he needed to transform a rye grass field into a wetland. He also received help and advice from the local office of the Natural Resources Conservation Service, the Oregon Department of Fish and Wildlife, and Ducks Unlimited. Last, but not least, he had the right site. "If you want to succeed in restoring a wetland," he says, "you need to know the site's opportunities and barriers, and then design with those in mind. I have the perfect conditions: a non-draining clay soil, flat topography, a large piece of ground, and minimal invasive plant contamination."

Diligent monitoring and maintenance of the site,

Flowers and grasses create a colorful tapestry on the Knaupp's wetland.

especially invasive plant removal, are critical elements of the project's success.

Mark cautions newcomers that, over time, the wetland takes on a life of its own. Mark's biggest complaint with the project has been the lack of flexibility of some of the Wetlands Reserve Program regulations. He has had some disagreements with program officials about site management and compatible uses. The murky wording of the compatible use regulations requires Mark to apply for permission to mow the native grasses even in September, in order to provide habitat for geese and promote growth of other native plant species. Twice a year, he and all the project partners walk through and evaluate the site, noting changes and needed adjustments.

On balance, he's grateful for the help from The Wetlands Reserve Program. "They allowed me to get my money out of the property and restore habitat at the same time. It's the best of both worlds for me."

"Buying the land, farming

it and then enrolling

in the WRP and restoring

it to wetland was a great

business decision."

EDITH LESLIE

Appreciation, respect and stewardship of this land,

moved from my past to my future. Who could wish for more?

EDITH LESLIE, 2001

TURNING LORE TO LEGACY

Family launches award-winning project

Coos Bay, Oregon

Edith and Willamar Leslie left California in 1970 to build a home at Beaver Hill, outside of Coos Bay. For years, as their children were growing up, the Leslies had taken family camping trips to this Oregon property. Retiring there proved irresistible.

Once settled, the Leslies raised a few head of cattle and sheep, then set out to replant trees. This piece of land, which has been in Edith's family for more than a century, is a centerpiece in the family legacy. Years ago, a cash-starved logger gave Edith's grandfather the land in exchange for an unpaid tab at the Prosper general store, which Edith's grandfather founded about 120 years ago. At the time, says Edith, the logged-over land was valued at less than 50 cents an acre — a mere fraction of the debt.

Edith's mother left that piece of land and the Coquille Valley when she was a young woman, but kept it alive in her memory. For years, the family heard stories of the soggy ground near Prosper. "Salmon so thick," relatives were told, "it almost seemed as if you could walk across the Coquille River on their backs."

Edith inherited the piece of squishy earth from her mother in the 1950s and continued to rent the pastures to a neighboring farmer for grazing. After her husband's death in 1992, neither Edith's two sons nor her only daughter had any interest in ranching the family parcel. For about five years they leased out the pastures for grazing. Edith loved that land, and wanted to find a way to transfer it to her children and grandchildren.

In her search for solutions, Edith met Michael Graybill, director of the South Slough National Estuarine Research Reserve. He explored with Edith and her children how they might restore the pasture to its natural state. Then, he connected the family with a cadre of scientists, engineers and ecology gurus at the South Coast Land Conservancy. They showed the Leslies how the family could create a conservation easement to keep the property.

This was no easy decision. Thirteen people representing three generations batted around the pros and cons of the easement, debating the trade-off between preserving the land and doing whatever they wished on it. Some family members balked at signing away forever the option of running livestock or harvesting hay. In 1997, they threw their hats together to establish a limited liability corporation, keeping the land in one piece they could all own.

The timing was good. In 1995, the state had new funding to restore wetlands and boost dwindling salmon populations. The South Coast Land Conservancy, in partnership with the Leslies, received a grant from the Oregon Governor's Watershed Enhancement Board to restore the hydrologic connections between the pastures and the Coquille River. The Natural Resources Conservation Service's Wetlands Reserve Program, United States Fish and Wildlife Service Partners in Wildlife Program, Oregon Department of Fish and Wildlife, Coquille Watershed Association and Ducks Unlimited also contributed to the effort.

The family received technical advice from many agencies and natural resource proponents. Some neighboring landowners, however, questioned the wisdom of converting farmlands to wetlands. One neighbor requested a dike around the project to protect his horse pasture from floods the restoration could cause. The Leslies agreed to his request, never considering they would have to flatten one of their favorite landmarks to get dike-building materials.

Red-legged frogs now thrive where cattle once fed.

When the restoration project was completed in October 1999, the whole Leslie family celebrated with an old-fashioned barbecue for neighbors, friends, and partners in the project. They thanked everyone who had worked on the project, and gave landowners a chance to explore and ask questions about the new landscape. Around plank picnic tables, Coquille Valley landowners, biologists and engineers exchanged ideas, which have spun off into a number of new private restoration projects.

Recognition for the project extended beyond the Leslies' community. The US Fish and Wildlife Service awarded the family the National Wetlands Conservation Award for the private sector.

But the real rewards lie among the coho salmon smolts darting in the upland stream, the ribbeting red-legged frogs, and many varieties of birds in the restored wetland's grassy areas. "Appreciation, respect and stewardship of this land," says Edith, "moved from my past to my future. Who could wish for more?"

Three Generations of the Leslie Family

Edith and Willamar Leslie

The Leslies worked together to restore their wetland on land that had been in the family for over 100 years.

Lael Leslie-Lepowski	Lorin Leslie	Lann Leslie
Jim Lepowski	Shannon Leslie	MelodyLeslie
Tris Lepowski	Forrest Leslie	Logan Leslie
Mei Lepowski	Lorelei Leslie	Megan Leslie

DOUG AND SHARON SINKO

If I had influence with the good fairy who is supposed to preside

over the christening of all children, I should ask that her gift to each child in the world

be a sense of wonder so indestructible that it would last throughout life.

RACHEL CARSON, 1992

EX-TEACHER COMES HOME

Childhood memories fuel arduous effort

Myrtle Point, Oregon

From the age of 12, Doug Sinko was raised on a Coquille River dairy farm. He spent his teenage years exploring the wetland: catching frogs, watching birds, becoming a "naturalist." In 1972, after five years of teaching, Doug returned to his roots and bought the 120-acre piece of land from his father. In 1979, he added the adjacent 240 acres.

Fourteen years later, Doug and his wife Sharon became the first organic dairy farmers in the Northwest. Soon after they received their organic certification, they cut their herd from 300 to 150, reduced the size of their pasture and decided to change the character of the retired pastureland. "Sharon and I are very excited about restoring the wetland," says Doug. "At times, the setbacks, delays and red tape test our commitment and patience. Finally, after 20 months, I think we have cleared all the hurdles."

Historically, Doug and Sharon's land near Myrtle Point was a willow-and-ash marsh. In the early 1900s, dairy farmers began settling along the major rivers in the Coquille Valley and changing the land. By the Depression, valley farmers had slashed and burned much of the vegetation to create pastureland.

In 1972, during the Sinkos' first year of the dairy operation, high water nearly gutted the river bank. That year, in their first restoration project, the Sinkos stabilized and reclaimed 1,000 feet of the bank. The bank had eroded roughly 40 feet each year, creating steep drop-offs in some places. With the help of the Soil Conservation Service (SCS), the Sinkos secured the proper permits – 13 in all – and sloped the bank, putting a rock toe at the bottom. They seeded and irrigated the slope in an effort to sprout a grass cover before winter. Other vegetation sprung up naturally. Now, large

Winter flooding and scouring creates this backwater on the Sinkos' wetland.

willows and alders shade a narrow, deep channel. "And," Doug adds, "the fishing is good."

The Sinkos attempted a second bank stabilization in 1992 — with different results. Again, they sloped back the bank. This time, upon the advice of the funding agencies, the Sinkos did not place rock at the toe of the slope. Doug and Sharon's fears were realized when, over time, the bank reverted to the wide, shallow channel it had been before the restoration.

"We haven't been able to establish a ground cover or trees," says Doug. "Any plants that we've added have been swept away by floods and erosion, just like we predicted."

In 1998, Doug and Sharon learned about the Wetlands Reserve Program (WRP), a federal program that compensates landowners in agricultural production for

"In reality, we are caretakers

of this earth for such a short time.

We would like to give

something back to the land

that has been good to us."

restoring and protecting their wetlands. The WRP offered the tools and incentives they needed to maximize the natural state of their wetland and pay off part of the land. The easement, which allows the owners to maintain control, was particularly attractive to them.

In 1999, after several brainstorming sessions with the South Coast Land Conservancy (SCLC) and Tom Purvis of NRCS on how to restore the wetland, Doug and Sharon applied to enroll 210 acres in the WRP. That's when the delays, frustrations and fun began. The Sinkos spent 18 months haggling with the Farm Service Agency (FSA) over the real value of the land. The SCLC negotiated, read over legal contracts and generally helped them navigate the arduous process. The Sinkos, with help from the SCLC, altered the standard WRP easement language. They were able to maximize the value of the land by relinquishing their rights to hunt, fish, control trespassers and mine below the surface of the land.

"Imagine our dismay," Doug says, "when two appraisals and four months later the Farm Service Agency required a third appraisal." The SCLC used grants from USFWS Coastal Wetlands grant and the Oregon Watershed Enhancement Board to compensate the Sinkos for the gap between what the WRP was willing to pay and the appraised value of the land.

Twenty months later, with all the paperwork complete and financial matters settled, Doug and Sharon look forward to working with SCLC and Ducks Unlimited to secure the necessary permits. They hope to begin construction in June 2002.

At times, they were fed-up and just wanted to walk away from the project. But then they remembered their greater purpose. "In reality," says Sharon, "we are caretakers of this earth for such a short time. We would like to give something back to land that has been good to us."

After a dry summer, the Sinkos still have water in a few spots on their Coquille Valley property.

TERRY WAHL

It may be those that do most, dream most.

STEPHEN LEACOCK
1869-1944

SHEEP FARMERS
SHIFT PRACTICES

Movable fence protects streams, fuels dreams

Langlois, Oregon

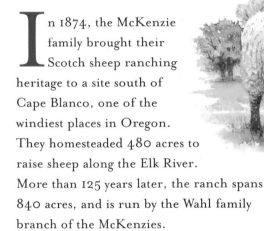

In 1874, the McKenzie family brought their Scotch sheep ranching heritage to a site south of Cape Blanco, one of the windiest places in Oregon. They homesteaded 480 acres to raise sheep along the Elk River. More than 125 years later, the ranch spans 840 acres, and is run by the Wahl family branch of the McKenzies.

Georgiana Wahl, her daughter Tooz and sons Terry, Bucky and Pete, tend to day-to-day operations on the South Coast ranch. Georgiana's other children raise sheep in various Willamette Valley locations. Family members share the management of the original ranch property, each having equal voice and power.

National trends, economic opportunities, and the conditions of the land helped shape the Wahls' ranch management techniques. In the 1960s and 70s, like many other ranchers, the Wahls removed willow, discouraged beaver, and cultivated as much land as possible. In the early 1980s, they relined the creeks with trees. "We planted trees on the land that wasn't really fit for grazing," says Terry, "and fenced to keep the livestock out of the creeks."

They also rotate the grazing areas on the ranch daily. "We practice intensive grazing — using movable electric fencing," says Terry. "We let them feed for a day and then move them to the next strip." This reduces the impact on the land and results in higher quality forage.

Since 1995, the Wahls have worked to restore streams, wetlands and fish with help from a long list of government programs. Harry Hoogesteger, coordinator of the South Coast Watershed Council, has cut some of the red tape of state and federal clean water and endangered fish recovery programs.

The Wahl family has raised sheep for over 125 years on their Cape Blanco ranch.

"Harry makes it all so easy," smiles Terry. "After a couple of years of working together, he has a good sense of our family operation and interests and has been able to match them with and help navigate us through a variety of government funding and permitting programs."

The Wahls have fenced and planted more than five miles of riparian area along the Elk River mainstem and its tributaries, to increase shade and keep livestock out of the streams. The fencing also has improved ranch management. "An example," says Terry, " is that when sheep get sick they go to wet places to die. The fencing eliminates the difficult task of hauling sick or dead sheep out of the wetlands."

In their restoration projects, the Wahls erect the fences and plant the trees, which include willow, spruce, shore pine, hemlock, cottonwood and western red cedar. The South Coast Watershed Council, Curry County Soil & Water District, Oregon Watershed Enhancement Board and the Environmental Protection Agency have contributed funding to the effort.

The rewards came quickly, according to Terry. Dense willow thickets and conifers sprang up along the river's edge, within the fenced-off seeps and in the wet areas around the ranch. Terry, a keen birder, notes, "It seems like the numbers and types of birds using the ranch have increased, though I haven't conducted a census to prove it."

One of the significant restoration projects took place on Cedar Creek, a tributary of the Elk River. A culvert there blocked fish passage upstream. The Wahls worked with a neighbor to replace the culvert with a rail car. This opened up a wetland and nearly a mile of salmon habitat. The South Coast Watershed Council helped coordinate the project and

". . . when sheep get sick they go to wet places to die. The fencing eliminates the difficult task of hauling sick or dead sheep out of the wetlands."

raise funds from the Oregon Wildlife Heritage Foundation, Oregon Watershed Enhancement Board, US Fish and Wildlife Service, South Coast Watershed Council and Curry County Soil & Water District.

The Wahls and the South Coast Watershed Council are now eyeing a new project on Swamp Creek, which also runs through the Wahl Ranch. Two impounded reservoirs there could provide ideal rearing habitat for coho salmon. The family plans to install fishways in both of the dikes that hold back irrigation water in order to open up nearly five acres of rearing habitat. The creek historically had runs of coho and chum salmon.

Once again, the South Coast Watershed Council is helping solicit project funds and in-kind services from US Fish and Wildlife Service, the South Coast Watershed Council, Curry County Soil & Water District and the Oregon Watershed Enhancement Board.

"Working with the government programs has been great for us," says Terry. "As long as they match our family management goals and strategies, we are open to new projects and partnerships."

Sheep, fish and wildlife live side by side on the Wahls' expansive ranch.

LOUIS AND MARR RANDALL

What is honored in a country will be cultivated there.

PLATO
347–427 BC

FARMER FINDS
NEW WAYS ON RIVER

Duck club proves profitable

Bonanza, Oregon

At age 13, Louis Randall moved from Fort Sumner, New Mexico to the Langell Valley, outside of Bonanza, Oregon – the backdrop for Zane Grey's book *Forlorn River*. After 58 years of ranching and farming, he is busy reclaiming over 1,000 acres of wetland on his 10,000-acre Circle 5 ranch.

Changes to the land, rivers and hydrology of the valley began in 1868, when the Langell family settled along the banks of the Lost River and rechanneled it to reclaim 4,000 acres of farmland. By 1904, as more families settled in the valley, the network of canals and irrigation ditches expanded in all directions, transforming arid sagelands to bountiful fields of grain and irrigated pasture. In 1902, the creation of the Bureau of Reclamation saw the beginning of more and larger diversion and irrigation projects in the Klamath Basin and Lost River Valley. The biggest change for the Circle 5

Ranch was the 1950s Bureau of Reclamation project that rechannelized and rerouted the Lost River through the swamplands to promote irrigation. The reconstruction dewatered the valley.

"We used to cut hay on the high ground, and then in 1945 we broke up 640 of the 2,000 acres of wetlands on the land," Randall says. He spent the summer of 1945 plowing and preparing the land. The following year, he planted it in oats. "It was all wetlands before the Bureau of Reclamation started moving things around," explains Randall.

"After a couple of years of good yields, the swamp land was never very productive," continues Randall. People had always hunted ducks on the ranch. "So in 1970, in order to make up for the economic loss of farming the marginal soggy ground, I decided to restore the wetland, improve the waterfowl habitat and start a hunting club."

"Anymore, a person can't change

or decide how to manage their

land without asking for permis-

sion, and the permission adds up

to a lot of paperwork permits and

multi-agency scrutiny."

So began his first restoration effort. He was a true pioneer, recreating the wetlands on his own long before any of the federal wetland enhancement programs existed. Using his own machinery, Randall dug a drain ditch for two center pivots and built a dike around the drainage ditch to keep the poorest soils wet.

"It was a lot easier back then. You didn't have to get everyone's permission and all kinds of permits to change around your land," Randall recalls. "Today it's a different story." As the habitat improved, the number of ducks and geese increased. He started up the club and charged people $10 to hunt in the wetland.

In 1989, he entered his first partnership with a federal agency, US Fish and Wildlife Service's Partners in Fish and Wildlife Program. He received funds for a 400-acre wetland restoration on the property near the Lost River. In 1996, he received additional funding to fence livestock from some spring-fed ponds.

Randall saw that traditional ranching needed to move forward and that the marshlands were

providing more income as a duck club than from crop production. He decided to put portions of the wetland into a permanent easement. In 1994, he enrolled 700 acres in the Wetlands Reserve Program and in 1997 added an additional 311 acres.

Although he was able to benefit from a variety of federal technical assistance and financial programs, the process was not without frustration. Randall is no stranger to water and natural resource management and policies. He has spent much of the past few decades sitting on the local Soil and Water Conservation District and irrigation boards. At one time, he was head of the Oregon Cattlemen's Association and a founding member of a group of ranchers interested in practicing ecologically sustainable ranching. All that experience with boards, bureaucracies and policies didn't prepare him for the exasperation of navigating federal grant and permit programs.

"Lucky for me, Jim Hainline of US Fish and Wildlife and a cast of others were there to shepherd me through the Corps of Engineers permits and archeological clearances," Randall sighs. He also had some differences of opinion with Natural Resources Conservation Service about active management activities such as mowing and burning. "Anymore, a person can't change or decide how to manage their land without asking for permission," Randall sighs, "and the permission adds up to a lot of paperwork, permits and multi-agency scrutiny."

In February 2001, he completed restoration and enhancement of 700 acres of moist soil wetlands, and 200 acres of adjacent uplands on land previously used for hay production and livestock grazing. Much of it took place on the land he had previously enrolled in the Wetlands Reserve Program.

The restoration funding was provided by the US Fish and Wildlife Service, Natural Resources Conservation Service, Ducks Unlimited and a North American Wetland Conservation Act grant. It supported construction of levees and water control structures to restore the historic hydrology in three independent wetland areas. The water control structures have also improved the capability to manage water on the existing wetlands.

Midway through the construction process, Randall had one more setback. Until he received a pond permit from Oregon Water

Louis Randall's pioneering efforts provide daily rewards. He began his restoration in 1970.

Resources Department, he could not direct water into the newly restored wetland. He was a bit confused about why he needed a permit. Prior to the restoration, the same land, then in farm fields, ponded water in much the same way. A year later, Randall and the wetland patiently wait for approval to let the water flow.

The Circle 5 Ranch shows how private landowners can integrate wildlife protection into agricultural operations. Thirty years of habitat restoration have attracted a variety of waterfowl, sandhill cranes, tricolored blackbirds, bald eagles and northern harriers to the wetland. "I'm proud of returning the land to how it looked when I moved here as a teen," Randall says.

JEAN AND JERRY HINES

Heroes are not grand statues framed against a red sky.

They are people who say this is my community

and it is my responsibility to make it better.

TOM McCALL
1913-1983

COUPLE LEARNS
FROM NEIGHBOR'S MISTAKE

Finds new bumps in road to restoration

Chiloquin, Oregon

Jerry and Jean Hines bought their 92 acres on the Sprague River, near Chiloquin, in 1995. Their goal for the riverfront property: creating a wildlife sanctuary. Their neighbor had already restored some wetland. "We figured we could learn from him before starting on our own," Jerry says.

The neighbor shared design and technique advice. But Jean and Jerry learned their biggest lesson from their neighbor's mistake. The Division of State Lands had stopped the project because the neighbor had not obtained the proper permits. The Hineses learned they needed state and federal permits before moving any dirt. They also learned from the neighbor about the US Fish and Wildlife Service's Partners in Wildlife Program and the Natural Resources Conservation Service's Wetlands Reserve Program, and pursued opportunities with both.

In 1999, the Hineses were ready to transform their ranch into a refuge for sandhill cranes, bald eagles, trout and other species they saw on the river, Jerry says.

"We learned," adds Jean, "that the path to restoring the historic wetland was neither straight nor smooth."

Philosophical differences, language barriers, human error and bureaucracy added twists to their bumpy restoration project. The Hineses also discovered the importance of wetland jargon. "We quickly learned from a very helpful US Fish and Wildlife Service employee," says Jean, "this restoration culture has its own vocabulary. We had to substitute 'berms' for 'levees' and 'marshes' for 'ponds.' The project goals had to include 'improving water quality' and 'fish habitat'."

The Hineses hoped to diversify habitat on their property to attract a variety of fish and wildlife. They planned to develop nesting

islands in the floodplains, create different water depths by enhancing the existing swales and complete a berm around the wetland to hold water. The Hineses submitted their first Division of State Lands permit application. But the state and federal agencies involved in the permit process reached an impasse. The Oregon Department of Fish and Wildlife and US Fish and Wildlife Service disagreed about fish entrapment and the best way to connect and restore floodplains and oxbows. "The Division of State Lands told us the two agencies had to work it out," Jerry says.

After months of discussion, they reached consensus when the Hineses proposed adding fish escape routes to each of the four fields they were turning into wetlands. They also planned

A winter snow and ice-coat blankets the Sprague River.

to deepen the oxbow between the outside bend and the confluences of the river so fish would not get trapped or isolated.

Finally, the Hineses learned they needed a separate permit for the oxbow work. The Division of State Lands is requiring them to apply for a general permit, rather than applying for a restoration and enhancement permit for construction of the berm connecting the islands to the rest of the ranch. Therefore, the application process will cost $225 and take longer for review. Perry Lummley, recently retired from the lands division, plans to help them through the permit process should they encounter more roadblocks.

"As a landowner, it would have

been helpful if there was one

point person, agreement

among the agencies, and a

knowledge of the true costs,

restrictions and timelines."

In the fall of 2000, the Hineses hit another bump. They had applied to enroll 80 acres into the Wetlands Reserve Program, but the agency neglected to file their application. The Hineses convinced agency officials to place their project on a waiting list, should funds become available. If so, the couple has a chance of being accepted in the 2002-2003 program. That is, if the program is funded.

Meanwhile, the Hineses await guideline development for the Natural Resources Conservation Service's Wildlife Habitat Incentives Program, which offers assistance to landowners interested in enhancing wildlife habitat.

Committed to their dreams for the wetland, the Hineses search for more funding. In early 2001, they applied for and were accepted into the US Fish and Wildlife Service's Partners in

Fish and Wildlife Program. In April, with permits in hand, Jerry began moving the earth. He created berms for the rice fields and the lower wetland restoration. He also prepared and planted the south barley field as a food plot for wild geese.

In another bump, the Oregon Watershed Enhancement Board (OWEB) in late spring denied the Hineses' request for funding. Thus, the couple is pursuing other avenues for funding that may be more receptive to their vision. They are optimistic about the Hatfield Fund, which provides funding for projects in the Klamath Basin.

The Hineses save time and money by doing their own surveying, design, drafting, engineering and earth-moving. But leaping over government hurdles is expensive and time consuming. "As a landowner," says Jerry, "it would have been helpful if there was one point-person, agreement among the agencies, and a knowledge of the true costs, restrictions, and timelines. Hopefully," he muses, "our experience can provide help for the next person, like our neighbor did for us."

Jerry Hines, sitting in his tractor, reflects upon his labor to restore historic wetlands and improve wildlife diversity in the Sprague River.

DAN AND KATHY RIDGEWAY

We have subdued the wilderness and made it ours. We have conquered the earth

and the richness thereof. We have indelibly stamped upon its face

the seal of our dominating will. Now, unlike Alexander sighing

for more worlds to conquer, we should address ourselves

to adding beauty to that glory and grandeur.

ALICE FOOTE MACDOUGALL

1867-1945

BUSINESS REALITIES
SPUR CHANGE

Neighbors object to government partnerships

Sprague River Valley, Oregon

The spotted owl was the first animal to change Dan and Kathy Ridgeway's life. Forest conservation practices to protect the endangered bird ended their Marion County business, Salem Lift Truck, Inc. They had sold new and used forklifts, and also had a rental fleet of forklifts. It was a very successful business until the spotted owl controversy destroyed the local timber industry. In 1994, they moved to the Sprague River Valley in South Central Oregon in search of a more peaceful life.

Dan's childhood memories shaped their dreams of a new life along the Sprague River. In his youth, his family had owned and managed an Idaho farm. "We thought we could make a go of a cattle operation in the Sprague River Valley," he says. But the cattle's impact on the river, combined with economic realities and riverbank reveries, gave the Ridgeways' life a new twist. "One might hear our story," says Dan, "and think we enjoy pain and suffering."

Before starting their new business in 1994, the business-savvy couple penciled out a plan for a cattle ranch that would turn a profit. During their first year in business, they leased 100 breed cows. The following year, they switched to running their own yearlings. The third year, they ran leased yearlings. The fourth and final year, they ran 200 leased yearlings and 100 pair of their own. Based on the numbers, their business appeared to be on track.

However, the Ridgeways noticed that the livestock degraded their property. Water levels rose and they were losing land to erosion. "Watching the water and land change," says Dan, "I realized that if I did nothing, I would find myself living on the banks of Lake Sprague." Again, Dan's memories of youth spurred a change for the couple. He

Instead of chasing cows,

building fences and managing

vegetation, Dan and Kathy

filled out forms, attended

meetings and spent hours

on the telephone.

remembered fishing for trout along lush streams: "I decided I wanted to do my part to restore the Sprague River."

In the final analysis, a decision to reduce the number of cattle would not make sense, so Dan and Kathy explored other ways to make money from their property. Recalling the advice of a neighbor, they applied to enroll their wetlands in the Wetlands Reserve Program (WRP). In 1998, the Ridgeways placed 259 of their 273 acres into the WRP. They used the money from the WRP to pay off the rest of the land. "At that point," says Dan, "I learned to walk in two very different worlds and develop both environmental and agricultural eyes."

Back at the drawing board, the Ridgeways researched ways to finance the restoration and identified potential new income sources. Instead of chasing cows, building fences and managing vegetation, Dan and Kathy filled out forms, attended meetings and spent hours on the telephone. The hardest part for the enthusiastic couple was waiting. "Once again,"

says Dan, "we created a business plan which has had to continuously float and change." This time, their plan for the property was a duck-hunting or fly-fishing club and an environmental destination for canoeists and birdwatchers.

"We entered the WRP with the expectation that NRCS would provide financial and technical assistance for our restoration project," says Dan. The couple, however, has not always seen eye-to-eye with the program administrators. After a year in limbo, they began looking for other potential funding sources.

Meanwhile, word of the Ridgeways' plans to restore the wetland filtered through the community. "Our neighbors began to shun us," says Dan. "One morning at the local restaurant we were seated with someone who spent the entire breakfast pounding the maple syrup pitcher on the table as he lectured us about the evils of government and taking lands out of production. We haven't returned there for breakfast."

Negativity toward the Ridgeways' new enterprise has been emotionally trying for the couple and has been a barrier to their search for funds. Many funding sources require support from the community or local watershed council. And the bureaucracies can be overwhelming.

Dan Ridgeway (left) paddles the Sprague River with a friend.

"Throughout this process," says Dan, "we have often felt lost, not knowing if we were at the beginning, midway through, or just about done and ready to construct. There are so many agencies, regulations and criteria that need to be met. From a landowner's perspective, they often do not seem well aligned, sometimes actually at odds."

Through hard work, persistence, and a few helpful folks, Dan and Kathy have gotten help. Curt Mullis and Faye Weekly of the US Fish and

Sandhill cranes are annual visitors to the Sprague River Valley.

Wildlife Service Ecosystem Restoration Office in Klamath Falls provided technical assistance, guidance and handholding as Dan and Kathy navigated through a maze of permit and grant applications. Oregon Watershed Enhancement Board and USFWS Jobs in the Woods Program have provided some funds.

The Ridgeways are optimistic that after four years of paperwork and negotiations, this summer they can finally send out invitations to the groundbreaking. In addition to adding 257 acres of fish and wildlife habitat, the Ridgeways hope their work will help simplify the process for other landowners who choose to restore the land. "Being pioneers," says Dan, "is never the easy road."

JACK SOUTHWORTH

A man is ethical only when life, as such, is sacred to him,

that of plants and animals as well as that of his fellow man,

and when he devotes himself helpfully

to all life that is in need of help.

ALBERT SCHWEITZER
1875-1965

Holistic Approach Inspires Change

Beaver abound in former pastureland

Seneca, Oregon

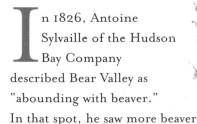

In 1826, Antoine Sylvaille of the Hudson Bay Company described Bear Valley as "abounding with beaver." In that spot, he saw more beaver than he had seen in any one place within the Northwest Territory. Now, 175 years later, Jack and Teresa Southworth restore the Silvies River on their Seneca ranch and learn patience from beaver, which once again are bountiful.

Jack's grandfather came to Bear Valley in 1885. He homesteaded 160 acres to raise hay for oxen to run a sawmill near Canyon City. He also set up Seneca's first sawmill, post office and general store. With profits from these enterprises, Jack's grandfather pulled together enough money to buy out some other homesteaders, expand the family holdings and start a cattle ranch.

The ranch, now 12,000 acres, continues to pass from generation to generation. Like his father, Jack experiments with the banks and streamside vegetation along the Silvies River. Many years ago, the family changed the land along the river from a dense willow thicket to hay and pasture land. Now, Jack and his wife Teresa are restoring the river banks to their natural, historic condition.

During one of the drought years of the 1930s, so the story goes, a friend of Jack's grandfather asked if the Silvies River still ran. "I can't tell," Jack's grandfather said. "It just kind of trickles from one beaver dam to the next." In 1955, the Soil and Water Conservation Service presented Jack's father with a plan to straighten the Silvies River through the ranch and build channels to improve irrigation of the adjacent pasture and meadow. "At the time, my father rejected that proposal," Jack recalls. "He subscribed to the graze-every-inch-of-ground fad of the time and fastidiously worked to remove the willows along the bank."

When Jack was 12, the family bought a new tractor and Jack got to drive it to the creek and pull out the last willow. "I doubt I will ever feel so good about anything again," Jack recalls, "I'd finally made our meadows look like the fields on the cover of *Successful Farming.* Surely I was doing the right thing."

By the early 1980s the ranch had gone from debt-free to incurring nearly $1 million in debt. In 1984, Jack and Teresa attended a week-long class on Holistic Management. "I was looking for any answer to help turn things around," he says. Holistic Management is a decision-making framework that assists farmers and others in establishing a long-term goal, a detailed financial plan, a biological plan for the landscape and a monitoring program to assess progress toward the goal. He walked away from the seminar with some good information on grazing and financial management practices, but tucked away in his back pocket, the center-piece goal-setting process. "It was as if I'd bought a new car but didn't have the key to start it," Jack says.

Three years later, Jack, Teresa, and their employee, Ed Newton, struggled through some goal-setting. They reflected on quality-of-life issues and identified what they were for the ranch. "It was a very difficult process for the three of us to describe quality of life," recalls Jack. "It was a whole lot easier to talk about calving or what fences needed fixing than to think about what you wanted out of life. In the end, we decided that we did not like to look out on a damaged creek." They felt the land needed some shrubs and a dense stand of perennial grasses. They also wanted willows along the stream banks to attract beaver and trout.

The goal-setting realigned their expectations. "Initially," says Jack, "we thought we always should try to hit a home run and make a lot of money. Once we acknowledged that we love where we live and what we do, we focused on hitting singles." They realized that financial success was only important if it brought about quality of life and environmental success as well. "The ironic thing is," Jack says, "is that we are better off financially as a result. We don't make a big profit, but we make a small one almost every year."

With their new vision, they plotted a new course for themselves. In the early 1990s, they received a grant from Oregon Department of Fish and Wildlife to fence the riparian portion of the ranch. The first grant covered a mile on each side of Silvies River. "It was pretty gratifying after the first-year willows and other wetland plants were visible," says Jack. In the second year, they received additional funding to construct two more miles of fencing on either side of the creek. "One thing that has made all this fencing possible is Brad Smith's ability at building fence," Jack adds. "Brad works with us here on the ranch and when Brad builds a rockjack, you know it is going to be there for a long, long time." The restoration

A willow grows from a rusty car. Nearly 50 years ago, Jack Southworth's father used the Chevy Impala to stabilize the bank.

has brought neither fortune nor debt to Jack and Teresa, but they feel proud of their healthy stream.

Two sandhill cranes call as Jack points out check dams and gravels they have added to the stream. In the restoration process, the Southworths debated over the depth of the river. They rejected a shallow, more erosion-prone channel, and refocused on their goal for the river: "It validated our choice to add the gravels and check dams to promote a higher water table, better diversity, a healthier meadow and better habitat for bass, suckers, and trout," Jack smiles. New sedges and wetland plants grow in the meadow each year. Plentiful beavers chew on the new willow plantings. The riparian zone may not have the willow lining they had envisioned, but the channel is definitely healthier. "We just have to learn to wait and let nature take her time," says Jack. "Beaver and

Wetland grasses and sedges take root in the once grazed floodplain of the Silvies River.

willows co-existed here for tens of thousands of years without our help."

The Southworths work with the Oregon Country Beef Program, which promotes ecological and sustainable land management practices. Their involvement brings them a higher price and more market certainty.

Teresa also raises llama and sheep. She cards, spins and felts the wool. Jack and Teresa are pleased with their decision for the family ranch. "I think it is a neat thing," says Jack, "when people can make a living off the land in an environmentally sound manner and support the community as well. That's what Teresa and I get to do."

DOUG MCDANIEL

When I am 85 years old and too old to ranch,

my dream is to be able to hobble down to the river

and catch a big trout . . . . I just hope the trout will still be around.

DOUG MCDANIEL, 2001

HURDLES
FRUSTRATE RANCHER

Red tape may squelch project

Lostine, Oregon

Doug McDaniel remembers the Wallowa River of his childhood, with its many meanders, log jams and places to fish and explore. Landowners along the bank had a long history of battling this river that encroached upon and stole their land.

After World War II, with access to larger tractors, landowners took control, straightened the river and moved its water through their land as quickly as possible. Over time, the loss of soil increased, rather than decreased, and the river required more maintenance than before. Stabilizing the banks to protect the land became a way of life. No one even considered restoring the twisty, old river.

Between 1960 and 1984, Doug ventured away from Wallowa County, started a successful logging enterprise and founded RD Mac Inc., a concrete company in La Grande.

In 1970, at the age of 35, Doug returned to Wallowa County and bought an overgrown, 600-acre ranch with a mile of Wallowa River frontage. Doug attacked the overgrown pastures to revive the farm. "I basically had to level it to re-create a 'normal' ranch. "It didn't take me very long to realize," he muses, "that it was going to be hard to make money off the 400 acres of upland pasture and 200 acres of bottomland."

He decided to restore the river to its historic shape. His first task was fencing off the river to protect it from cattle. Willows and other trees stabilized the banks and shaded the river, but high water continued to scour and carry away soil, woody debris and other nutrients important to fish. In March 1996, Doug partnered with Oregon Department of Fish and Wildlife (ODFW) to construct a fish screen on the Cross Country Ditch. He worked with them on installation of a second fish screen. In 1999, he constructed a fish ladder up to the pond.

Aerial photographs and walks through bottomlands allowed Doug to trace the meanderings of the Wallowa River as it once was. When the former cement company operator penciled out what it would cost to extract and haul the gravel to re-create the old channel, the math looked good: "I knew the sale of the gravel would come close to paying for most of the project expenses."

Doug set up visits with resource agency staff to brainstorm ways to improve fish habitat. ODFW biologists came up with some concepts. Other agencies took an interest in the project, and began to amass funding. It seemed like a model project: a willing landowner with a large section of the river and adjacent bottomland willing to fund the restoration. Between 1992 and 1994, after countless meetings, discussions, and site visits, Doug's patience began to wither. The delays and lack of decision blocked his dream for five years.

In 1999, Doug tried again, enlisting help from Wallowa Resources to revive the project and secure the proper permits. Wallowa

Doug McDaniel created this slack water pond beside the Wallowa River – the first phase of his restoration.

Resources is a non-profit that promotes the importance of forest and watershed health to community well-being. Doug again weathered site visits, meetings about design and negotiations. To comply with permit and grant funding requirements, he hired an engineer to do a hydrologic analysis. Twelve thousand dollars later, the design "controlled" the river, rather than restore it to its historic channel.

Doug knew the plan wouldn't work. For more than 60 years, he had seen the Wallowa River shrink, swell and spread. But he kept his mouth shut and deferred to "the experts."

"Much to everyone's relief," says Doug, "one person had the courage to say the plan sucked. Once again," he says, rolling his eyes, "my original plan started to look pretty good."

Still, no person or agency took the lead. "Each time I turned around," says Doug, "there were more problems and delays with permits and the price tag kept going up."

Perhaps it's his temper, his straight-forward nature or his impatience with inefficiency

Trees shade the Wallowa River, cooling the water for fish.

that have delayed the project and prevented him from restoring his childhood fishing hole, he says. "However, coordination of all these 'experts' and permit and bureaucratic snafus have added their share to the delays."

The willows and other vegetation along the stream bank continue to grow taller and shade the stream. Doug feels the bureaucratic hurdles likewise grow. Once again, he says he considers walking away from the project.

"When I am 85 years old and too old to ranch, my dream is to be able to hobble down to the river and catch a big trout. I have my doubts," he sighs, "that this project will come to fruition. I just hope the trout will still be around."

At publication deadline, The Wetlands Conservancy and Wallowa Resources were working with Doug to restore his stretch of the Wallowa River.

MARK TIPPERMAN AND LORNA WILLIAMSON

Each time we learn how to join together and mend our ties

with our own little place called home, we link our souls with the soils

that sustain us, and nurture the network that is shaking the earth."

ELAN SHAPIRO, 1992

VISIBLE PROJECT
BOASTS MANY PARTNERS

Urban refugees see instant results

Starkey, Oregon

The traffic and sprawl in Snohomish County, Washington sent Mark Tipperman and Lorna Williamson in search of a quieter, more remote setting. After searching for 400 acres or so east of the mountains, they landed a 2,500-acre ranch along McCoy and Meadow creeks in Oregon's Blue Mountains.

When they bought the Starkey, Oregon property in 1990, Mark and Lorna planned to run some cattle. What they really wanted, however, was to return the land to a more natural state. They thinned some timber and replanted some trees, but McCoy Creek and the meadows surrounding it became the centerpiece for what was to become Eastern Oregon's most visible wetland enhancement project. "I think just about every federal and state resource agency has participated in this project," says Mark.

"We can honestly say," adds Lorna, "that we count the people we have worked with as friends. There have been a lot of them."

First, Oregon Department of Fish and Wildlife (ODFW) and the Bonneville Power Administration (BPA) helped them enlarge an exclosure to keep the cows away from McCoy Creek. Instead of the 20-foot corridor constructed by the previous owner, the new fenced-in exclosure protected the creek and the adjacent meadow. Others noticed — and replicated — the Tippermans' early success. "Our project," says Mark, "has inspired other landowners in the area to fence their riparian areas."

Members of the Confederated Tribes of the Umatilla Indian Reservation viewed the Tipperman ranch as a worthy project. The site is believed to have been the largest summer tribal encampment in the Grande Ronde region. The 10 tribes encouraged Mark and Lorna to let the creek wander the way it used to, says Mark, an idea initiated and supported

One year after project completion, McCoy Creek showed most of the characteristics of a stream reach in good condiditon.

by the Natural Resources Conservation Service (NRCS).

Mark and Lorna enrolled 500 acres into the Wetlands Reserve Program (WRP). They paid off the land debt with the WRP compensation. A lawyer seasoned in land use and real estate law, Mark secured tax deductions for the added values he donated to the WRP easement in the form of timber. He also worked with ODFW to develop and sign off on a wildlife management plan that allowed the land to stay in farm-use deferral, so he would not lose his farm tax benefit by establishing the WRP conservation easement.

For the past five years, Mark and Lorna have relied on technical expertise from agency staff in restoring McCoy Creek's flow through the marshy vegetation. "It's worked well," says Mark. "They all discuss and argue and come up with alternatives. Then, we make the final decision, usually by consensus."

Goals for the project reflect the diversity of partners involved. Mark and Lorna strive to improve fish habitat, protect beavers, restore

"We can honestly say that we count the people we have worked with as friends. There have been a lot of them."

native meadow and grassland plants, control noxious weeds, boost water quality, reconnect the stream to the floodplain, improve groundwater input and restore the straightened channel into a meandering creek.

With input from a host of agencies, this restoration project and other conservation practices have expanded to cover virtually the entire 2,500 acres. While the wide range of opinions has strengthened the project design, it also has multiplied the amount of time spent on technical assessments and planning. A private contractor and the NRCS worked on the stream channel morphology and hydrology. The Environmental Protection Agency worked on permitting. ODFW scrutinized fish and wildlife ecology and habitat needs, and constructed and maintained the riparian fence.

Oregon Department of Environmental Quality monitors water quality. The Umatilla Tribe has coordinated and managed the overall project and supplied biological and other analyses. The alphabet soup of funding sources includes Oregon Department of Environmental Quality, US Fish and Wildlife, US Environmental Protection and Natural Resources Conservation, the tribes and BPA. Results have been immediate. "Within one month of the phase one construction," says Lorna, "native sedges and rushes had re-established, and aquatic macro-invertebrates had moved in bigtime." Meanwhile, she continues the ongoing task of noxious weeds eradication.

In August 2001, a new span bridge was under construction to replace an old collapsed culvert. The fish-friendly bridge design and construction added Union County and the

US Forest Service to their long list of partners. By summer 2002, phase 2 of the project will be complete and water will be released into the final section of the recreated meandering channel.

Mark reflects upon their good fortune and the tremendous help they receive through their many partnerships: "We feel lucky and proud to be the hosts and particpants in the restoration of this section of McCoy Creek."

Beaver manipulate water levels and steer the course of the river.

This fish habitat log on McCoy Creek was placed by the Confederated Tribes, one of the many partners in the Tippermans' project.

CONSERVING WETLANDS

Wetlands are among the most important ecosystems on earth. They filter the landscape: purifying polluted rivers, preventing and minimizing flooding, protecting shorelines and replenishing groundwater. ◆ Historically, people regarded wetlands as "wastelands," barriers to development, and breeding grounds for mosquitoes, insects and disease. Considered useless, wetlands were too shallow to swim in and too wet to farm. People literally and figuratively paved them over for other uses. ◆ In just over two centuries, development has gutted many wetlands. In 1789, about 221 million acres of wetlands covered the lower 48 states, according to the US Fish and Wildlife Service. Now, this area includes 104 million acres of wetlands — less than half of the original acreage. ◆ The Oregon statistics also are staggering. Agriculture, commerce and other developments supplant many acres of former wetlands. Statewide, about 1.4 million acres of wetlands remain — 2 percent of Oregon's total land surface. In the Willamette Valley alone, more than 500 acres of wetlands are lost each year, according to the Oregon State of the Environment Report 2000. ◆ Shorebirds, waterfowl, fish and other wildlife depend on wetlands for survival. ◆ Nationally, 35 percent of all rare and endangered species depend on wetlands. As wetland habitat is destroyed, the number of species threatened with extinction increases. ◆ Gone are many of the species that inhabited these lost wetlands. This elevates the importance of the remaining wetlands, often shifting longstanding patterns. When a wetland is destroyed, for instance, migrating birds may be forced to change traditional migration routes. Similarly, other species must adapt or die. ◆ Even as our appreciation of wetlands grows, they are filled, dredged and drained. Private landowners, often in a strong position to restore and conserve wetlands, are choosing to do so. In conserving wetlands, they can realize financial benefits, including direct income, estate tax reductions and, in some cases, income and property tax reductions. ◆ Landowners have reasons beyond financial incentives to protect wetlands. Some farmers who use flooding as part of their operations improve topsoil retention, accelerate the breakdown of crop residue, decrease weed growth, diminish their need for fertilizers and reduce flood impacts to their agricultural land. Ranchers prevent loss of range land by restoring eroding stream banks, providing alternate watering sites for their animals and fencing wet areas on their property. For those who appreciate wildlife, wetlands that support waterfowl and shorebirds provide spectacular displays. ◆ A growing number of programs help landowners conserve and restore wetlands. Some restoration methods have been used for decades. Others are new. Choosing the right approach depends on a landowner's vision and needs, the wetland's functions and how the particular wetland fits into the larger landscape.

Restoring a Wetland

To begin planning a wetland restoration project, reflect and assess your situation. The clearer the priorities for the land, yourself and the rest of your family, the easier it is to develop a plan.

Start by asking: What is important? Improving wildlife habitat? Or assuring that heirs inherit a livable piece of land that is not a tax burden?

Ask yourself:

◆ What is the site's potential for restoration?

◆ What are the limitations of the land?

◆ What are my land management goals for now and for the future?

◆ What are my personal and family opportunities and limitations?

◆ What financial needs do we anticipate now and in the future?

The answers to these questions can lead to a strategy. Further information about community resources can help design that strategy:

◆ What kinds of technical assistance will help reach my goals?

◆ Who else can help?

◆ How can I include these experts in this process?

The clearer you are on what you want to do and what role you want to play in the process, the easier it is to elicit help from other individuals, groups or agencies.

There are many options. You can tailor your strategy to your particular property and circumstances. You can sell or donate the land, lease it or will it to someone. You can develop parts of the land. You can transfer certain property rights and responsibilities to a group, agency or individual through an easement or management agreement. You can manage it yourself or ask someone else to manage it. Common options, with considerations for landowners, are listed in the chart on the following page.

CHOOSING THE BEST OPTIONS

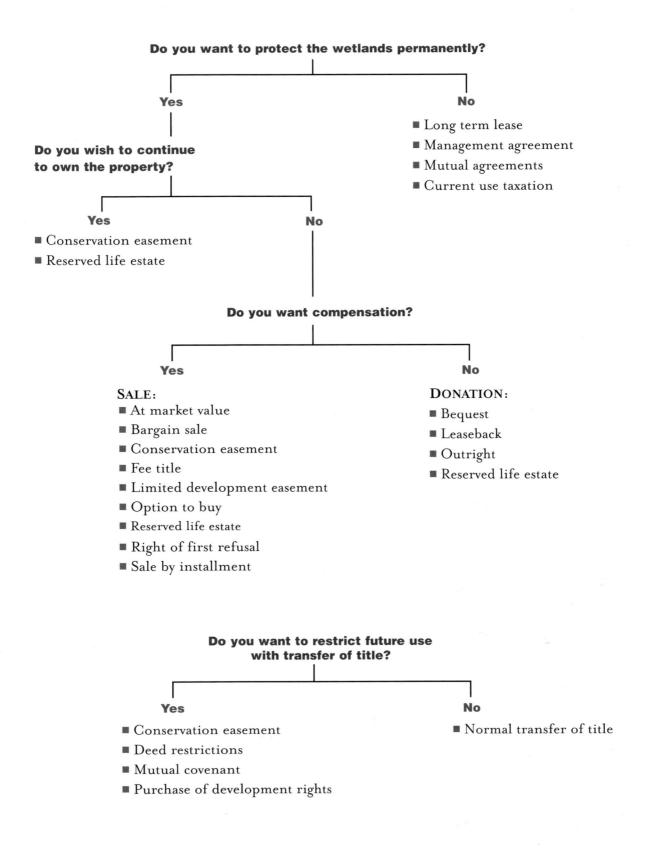

Do you want to protect the wetlands permanently?

Yes

No
- Long term lease
- Management agreement
- Mutual agreements
- Current use taxation

Do you wish to continue to own the property?

Yes
- Conservation easement
- Reserved life estate

No

Do you want compensation?

Yes

SALE:
- At market value
- Bargain sale
- Conservation easement
- Fee title
- Limited development easement
- Option to buy
- Reserved life estate
- Right of first refusal
- Sale by installment

No

DONATION:
- Bequest
- Leaseback
- Outright
- Reserved life estate

Do you want to restrict future use with transfer of title?

Yes
- Conservation easement
- Deed restrictions
- Mutual covenant
- Purchase of development rights

No
- Normal transfer of title

APPROACHES TO
LAND CONSERVATION AND PROTECTION

The following section explains techniques landowners use to conserve and protect wetlands. People generally sift through options that fall under three broad categories based on whether they wish to:

1) Maintain or own the property

2) Permanently transfer the title in exchange for payment

3) Transfer the title without compensation.

MAINTAINING PROPERTY

CONSERVATION EASEMENTS

Landowners can restrict how land may be used through written agreements, called easements. These become part of the property deed and stay with the land, binding subsequent property owners to the terms of the agreement.

In conservation easements, a landowner retains title to property, but transfers certain property rights to a land trust, government agency or nonprofit conservation organization. Through the easement, the landowner can restrict the type and amount of development on a piece of property in order to protect significant natural features, including wildlife or habitat.

Each conservation easement is tailored to the particular piece of property and the wishes of the landowner. The parties involved can renegotiate the easement if circumstances change.

Advantages
- Easements provide income tax, estate tax and gift tax benefits if the easement is donated or sold at less than market value.

- The property owner retains ownership of the property while potentially receiving income tax, estate tax and property tax reductions.

Disadvantages
- Easements can involve giving up some property usage rights.

- The landowner maintains the land and is responsible for expenses, including taxes.

LEASES

A landowner who is not in a position to manage the wetland as required by a conservation easement can rent the property to a land trust, conservation organization or government agency. Under this option, the landowner can require a tenant to manage the property for a specified period of time. Landowners can structure the lease with or without rental payments.

Advantages

◆ The landowner can receive payments for the leased property.

◆ The landowner can protect the land for a specified period without transferring the land to another entity.

◆ The landowner can terminate the lease if the property is not being used as directed.

Disadvantages

◆ Leases expire.

Unless provisions are made by the landowner, leases generally allow unrestricted and exclusive control of the land by the organization or agency leasing the property.

MANAGEMENT AGREEMENTS

Landowners can establish a management agreement with a land trust, conservation organization or government agency.

Advantages

◆ The landowner may be able to receive direct payments or other types of financial assistance.

◆ The landowner can often use the services of the land trust, conservation organization or agency to develop a site management and maintenance plan.

◆ It is easier to terminate a lease with a management agreement than with some other arrangements.

◆ Payments or cost-share maybe available for management or maintenance activities.

Disadvantages

◆ Management agreements expire.

RESTORATION

Landowners reintroduce water and native vegetation to restore the natural elements of the wetland.

Advantages

◆ Technical and financial assistance is available for many if not all project expenses.

◆ Restored wetlands can create views and attract wildlife that boost property values and quality of life.

Disadvantages

◆ Restoration without outside financial assistance can be expensive.

◆ Restoration is not always entirely successful.

◆ Restoration and rehabilitation of a site is generally a long-term commitment.

LIMITED DEVELOPMENT STRATEGIES

A landowner can restrict development to the "least environmentally significant" portions of property and use the proceeds to finance conservation on the remaining land or for other purposes.

Advantages

◆ A landowner can raise the money necessary to protect the more environmentally sensitive property.

◆ A landowner can combine conservation and limited development to help meet financial needs.

◆ A landowner can realize tax advantages by recording an easement over the undeveloped part of the land.

Disadvantages

◆ It can be difficult to determine which areas of the property are the least sensitive.

◆ Limiting land development can reduce its profitability.

◆ Adjacent land use may affect the wetland area.

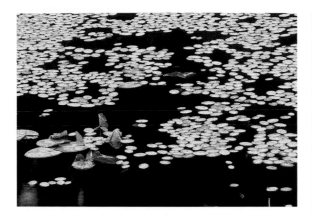

TRANSFERRING PROPERTY

REMAINDER INTERESTS

A landowner can postpone the transfer of property until after his or her death or after the death of subsequent owners. Through this arrangement, called remainder interest, the landowner can sell or donate property to a land trust or other nonprofit conservation organization.

Advantages

◆ Landowners enjoy all rights to the property during their lifetime, except those that degrade the natural resource value.

◆ Landowners provide future protection of the property.

◆ Donation for conservation purposes qualifies the landowner for a tax deduction, discounted in proportion to the anticipated length of time before the grantee takes over the interest.

◆ Whether the land is sold or donated, dedication of the remainder interest reduces the burden of the estate taxes.

Disadvantages

◆ The designation may restrict some uses of the land during the landowner's lifetime that may degrade the natural resource value.

PERMANENT TITLE TRANSFER WITH COMPENSATION

Sale Option

Landowners can choose from these four sale options:

1. **Fair market value**: The landowner sells the property for its fair market value.

2. **Bargain sale**: The landowner sells the property to a land trust, conservation organization or agency at a price below the fair market value. The difference between the sale price and fair market value can be characterized as a donation.

3. **Installment sale**: The landowner sells the property to a land trust or conservation organization where all or part of the consideration is deferred and paid in successive years.

4. **Right of first refusal**: The landowner gives a land trust or conservation organization the option to match a purchase offer and acquire the land if another buyer approaches the landowner.

Advantages

◆ Sale at full market value allows the landowner to receive full value for property.

◆ Bargain sales offer a tax deduction and reduction of capital gains taxes to the landowner.

◆ Installment sales can defer the actual capital gains tax until the purchase with which to pay the tax is in hand.

◆ Right of first refusal can give land trusts and other conservation organizations time to acquire the funds necessary for purchasing the land.

Disadvantages

◆ Most land trusts and conservation organizations have limited budgets and can rarely pay full market value for wetlands.

◆ If the land value has appreciated since it was purchased, the landowner becomes liable for the income tax on the capital gain.

TITLE TRANSFER WITHOUT COMPENSATION

Donation of Land

Landowners can choose from three types of donations:

1. **Outright donation** grants full title and ownership to the conservation organization, community or government agency receiving the donated property.

2. **Donation by deathtime** transfers property through a will.

3. **Donation with a reserved life estate** permits the landowner to use the donated property during his or her lifetime and the lifetimes of designated family members.

Advantages

◆ Donation provides total protection for a wetland.

◆ Landowners can receive income tax deductions and possible estate, gift and property tax breaks.

◆ Land trusts and conservation organizations, which may not have the budget to buy wetlands, can fulfill their mission to protect wetlands.

◆ Outright donation requires little negotiation and can be completed quickly.

◆ Donation at deathtime allows the landowner and their family to retain interim control and full use of their property, while ensuring protection after the landowners' death.

◆ Donation with reserved life estate allows the landowners and their family to continue to live on the land, while ensuring it future protection.

Disadvantages

◆ The landowner forfeits potential income from the sale of the land.

◆ Maintenance and other associated costs taken on by the land trust or organization may be more costly than a conservation easement.

◆ There is no income tax deduction for a donation by deathtime transfer.

◆ The landowner is responsible for property taxes as long as they remain in possession of the land.

◆ Many land trusts may not be able to accept the donation without additional funding for an endowment to support long-term management of the property.

Tax relief from donation with a reserved life estate generally applies to farms and personal residences, and in some cases wetlands may not qualify.

GRANT AND TECHNICAL ASSISTANCE PROGRAMS FOR WETLANDS RESTORATION

There are many state and federal programs that provide financial, technical and advisory assistance to landowners. The Grant and Technical Assistance Programs for Wetlands Restoration Chart on page 67 summarizes the elements of each of the state and federal programs. It can help narrow down which programs might meet your needs. Once you have identified a likely match, refer to the program profiles which follow for more details. Then select those that fit. The preceding stories illustrate the use of these programs in projects.

SUMMARY OF GRANT AND TECHNICAL ASSISTANCE PROGRAMS FOR WETLANDS RESTORATION

Program	Financial Assistance	Technical Assistance	Grant Portion of Total Cost	Target Species	Program Affiliation
Access Habitat Program	✓	✓	75%	Wildlife	State
Conservation Reserve Enhancement Program	✓	✓	Greater than 75%	Fish	Federal State
Conservation Reserve Program	✓	✓	25-50%	Fish and wildlife	Federal
Emergency Conservation Program	✓	✓	Greater than 75%	Other	Federal
Environmental Quality Incentives Program	✓	✓	50-75%	Other	Federal
Farm Credit Programs	✓			Other	Federal
Jobs in the Woods	✓	✓	Greater than 75%	Fish and wildlife	Federal
North American Wetlands Conservation Act	✓		25-50%	Wildlife	Federal
Oregon 25% Tax Credit	✓		25%	Fish	Federal
Oregon Watershed Enhancement Board	✓		25-75%	Fish and wildlife	State
Partners for Fish and Wildlife	✓	✓	25-50%	Fish and wildlife	Federal
Wetlands Reserve Program	✓	✓	50-75%	Fish and wildlife	Federal
Wildlife Habitat Incentives Program	✓	✓	50-75%	Fish and wildlife	Federal

GOVERNMENTAL TECHNICAL ASSISTANCE

ACCESS AND HABITAT PROGRAM (AHP)

Landowners with projects that have potential benefit to wildlife habitat or that increase public hunting access on private land can qualify for Oregon Department of Fish and Wildlife Access and Habitat funding. Projects might include: improving vegetation on wild lands, developing water in arid regions, reclaiming habitat by vehicular restrictions or fencing to control movements of wildlife or livestock. Projects can be on public or private land.

The Access and Habitat Program board pays particular attention to projects that reduce economic loss to landowners and those which involve funding commitments from other organizations and agencies. In-kind contributions of labor, equipment and material are also viewed positively.

For more information, contact your local Oregon Department of Fish and Wildlife Office. Information can also be obtained at: www.dfw.state.or.us/ODFWhtml/Wildlife/ahpgm.html

THE CONSERVATION RESERVE ENHANCEMENT PROGRAM (CREP)

The Conservation Reserve Enhancement Program (CREP) is a voluntary program that can provide farmers and ranchers financial incentives, over a period of 10 to 15 years, for removing lands from agricultural production. This joint federal and state program targets significant environmental effects resulting from agriculture. The Oregon CREP was developed to help restore habitat for endangered salmon and trout. The program hopes to restore freshwater riparian habitat along as many as 4,000 miles of stream banks throughout the state.

Program Goals

◆ Reducing water temperature to natural ambient conditions.

◆ Reducing sediment and nutrient pollution from agricultural lands adjacent to streams by more than 50 percent.

◆ Stabilizing the banks along critical salmon and trout streams.

◆ Restoring stream flow and land formations to their natural state.

Oregon CREP provides landowners wishing to improve conservation practices with four possibilities: annual rent, maintenance incentives, cost-sharing and an incentive for reducing the cumulative impact on the environment.

The maximum annual rental payment is $50,000 per person per year and cannot be higher than local rents for comparable land.

For more information, contact your local USDA Service Center, Soil and Water Conservation District office or the Oregon Watershed Enhancement Board. Information can also be obtained from the Farm Service Agency (FSA) web site at www.fsa.usda.gov/dafp/cepd/crpinfo.html.

CONSERVATION RESERVE PROGRAM (CRP)

The Conservation Reserve Program rents property from eligible landowners who agree to take environmentally sensitive farmland out of agricultural production. The agency shares the cost of the materials, labor and equipment landowners use to establish protective cover on their property. The program is designed to protect environmentally sensitive farmland from erosion, improve water quality, reduce surplus farm commodities and enhance wildlife habitat.

The maximum annual rental payment is $50,000 per person per year and cannot be higher than local rents for comparable land.

The program is administered by the United States Department of Agriculture. For more information, contact the local Conservation District, watershed council, Natural Resources Conservation Service office or Farm Service Office. Information can also be obtained from the FSA web site at www.fsa.usda.gov/dafp/cepd/crpinfo.html

EMERGENCY WATERSHED PROTECTION (EWP)

Landowners can receive financial and technical assistance to restore watershed areas that have been damaged by floods, fire, drought or other natural occurrences. The Emergency Watershed Protection Program buys floodplain easements and helps landowners with activities that repair conservation practices, remove debris from streams, protect destabilized streambanks and establish cover on critically eroding lands.

The program objective is to protect people from the imminent hazards caused by natural disasters.Thus, people are eligible for assistance even when there is not a national emergency declared.

For more information, contact your local Soil and Water Conservation District, watershed council or Farm Service Agency office. Information can also be obtained from the FSA web site at www.fsa.usda.gov/pas/disaster/ecp.htm

ENVIRONMENTAL QUALITY INCENTIVES PROGRAM (EQIP)

Commercial farmers and ranchers can solve point and non-point source pollution problems through technical, financial and educational assistance from the Environmental Quality Incentives Program. Eligible agricultural producers work on five- to 10-year contracts to establish permanent vegetative cover, retain sediment and stabilize water control structures. The program may share the cost of terraces, filter strips, tree planting, animal waste management facilities and permanent wildlife habitat. Also, incentive payments may be available for land management practices, such as nutrient management, pest management and grazing land management.

Fifty percent of the funding available for the program is targeted at natural resource concerns relating to livestock production. The program is jointly administered by the Natural Resources Conservation Service and the Farm Service Agency.

For more information, contact USDA, Natural Resources Conservation Service. Information can also be obtained from the NRCS web site at www.nhq.nrcs.usda.gov/OPA/F960PA/equipfact.html

FARM SERVICE AGENCY (FSA) FARM CREDIT PROGRAMS

Landowners with Farm Service Agency loans may consider three programs in which the US Fish and Wildlife Service provides technical assistance in conserving wetlands.

Two of these programs involve disposal of inventory farm property obtained through loan failure. The Service reviews these inventory properties and recommends:

- Perpetual conservation easements to protect and restore wetlands and conserve other important natural resources and
- Fee title transfer of inventory properties to state or federal agencies for conservation purposes.

The third area in which the Farm Service Agency provides technical assistance involves property owned by its borrowers. The agency evaluates the natural resource values of property secured through FSA loans and recommends contracts in which borrowers voluntarily set aside lands for conservation in exchange for partial debt cancellation.

The FSA is the primary manager of inventory easements, and receives approximately 40 percent of the fee title transfers. These lands become part of the refuge system. In addition, the FSA restores wetlands and other important habitats on FSA easements and transfer properties.

For more information, contact your local Soil and Water Conservation District, Watershed Council or Farm Service Agency office. Information can also be obtained from the FSA web site at: www.fsa.usda.gov/dafp/cepd/crpinfo.htm

JOBS IN THE WOODS (JIW)

The Jobs in The Woods Program is administered through the Oregon Field Office of the US Fish and Wildlife Service. The program goal is to restore watershed functions and processes in key watersheds, while providing local employment to communities with dislocated forest industry workers. The program funds restoration activities that correct the causes of watershed degradation. The objectives are long-term, sustainable solutions rather than short-term fixes.

The program focuses on adjoining parcels to maximize the positive impacts of watershed restoration.

For more information, contact the Oregon Field Office, US Fish and Wildlife Service in Portland at 503.231.6179 or Klamath Basin Ecosystem Restoration Office at 541.885.8481. Information can also be obtained from the US Fish and Wildlife Service website at www.r.1.fws.gov/jobs/ index.htm

NORTH AMERICAN WETLANDS CONSERVATION GRANT PROGRAM (NAWCA)

Landowners interested in acquiring, restoring, enhancing, managing and creating wetland ecosystems are eligible for funds through the North American Wetlands Conservation Act. The program encourages public-private and state-federal partnerships, with a strong interest in wetland habitats for migratory birds. The landowner or other non-federal partner must provide at least a 50 percent match for both the small grant (up to $50,000) and large grant programs (up to $1 million). The application process is complex and there is a high degree of national competition. A significant amount of lead time, pre-planning and advanced commitment of funding by project partners is required.

For more information, contact a local US Fish and Wildlife Service Office. Information can also be obtained at: www.northamerican.fws.gov/nawcahp.html

OREGON 25% TAX CREDIT FOR FISH HABITAT IMPROVEMENT

Landowners interested in improving fish habitat and preventing the loss of fish in irrigation canals by installing fish screens can qualify for tax credits. These amount to 25 percent of the cost for voluntary fish habitat improvements and 50 percent of the cost for required fish screening, bypass or fishway devices. All projects must be pre-certified by Oregon Department of Fish and Wildlife.

For more information contact a local Oregon Department of Fish and Wildlife office. Information can also be obtained at www.leg.state.or.us/ors/315.html.

OREGON WATERSHED ENHANCEMENT PROGRAM (OWEB)

Landowners are eligible for funds to enhance and manage riparian and associated upland areas to improve water quality. Funds also may be used to benefit fish and wildlife. Program funds help implement the Oregon Plan for Salmon and Watersheds.

Landowners or other partners must supply a minimum of 25 percent cost share. Landowners must agree to secure all the necessary permits, continue maintenance of the land and write monitoring reports.

For more information, contact your local Conservation District, watershed council or the Oregon Watershed Enhancement Board office in Salem at 503.986.0178. Information can also be obtained at www.oweb.state.or.us

PARTNERS FOR FISH AND WILDLIFE (PFW)

Private landowners who wish to restore, enhance and manage riparian, wetland, instream and upland habitats can receive technical and financial assistance from the US Fish and Wildlife Service's Partners for Fish and Wildlife Program. The program emphasizes the reestablishment of native vegetation and ecological communities for the benefit of fish and wildlife in concert with the needs and desires of private landowners. Projects must provide benefits to federally threatened and endangered species or species of concern, depleted native fisheries, neotropical migrant birds, waterfowl or the National Wildlife Refuge System. Project contributions are typically limited to 50 percent of project costs.

US Fish and Wildlife Service staff may advise landowners on the design and location of potential restoration projects. They also may design and fund the projects themselves under a voluntary cooperative agreement with the landowner. Under such agreements, the landowner maintains the restoration project for at least 10 years.

For more information, contact the Oregon Field Office of the US Fish and Wildlife Service 503.231.6179. Information can also be obtained from the US Fish and Wildlife Service web site at www.r1.fws.gov/oregon/hcr/pffw.htm

WETLANDS RESERVE PROGRAM (WRP)

Landowners with wetlands in agricultural production can receive payments for restoring and protecting their wetlands through the Wetlands Reserve Program. The program shares the cost of habitat restoration with landowners. It also pays landowners as much as the agricultural value of land for granting a conservation easement and maintaining wetland values. Easements and restoration cost-share agreements establish wetland protection and restoration as the primary land use for the

duration of the easement or agreement. In all instances, landowners continue to control access to their land.

The program is offered to high-priority sites, which have the potential to contribute to desired ecosystem functions. These include:

- ◆ Agricultural land with restorable wetlands

- ◆ Former or degraded wetlands occurring in pasture, range or forest production lands

- ◆ Riparian areas that connect with protected wetlands, along streams or other waterways

- ◆ Wetlands previously restored by an individual or under another federal or state program that are not protected by long-term easement.

For more information, contact your local Natural Resources Conservation Service office. Information can also be obtained from the NRCS web site at: www.fb-net.org/wrp.html

WILDLIFE HABITAT INCENTIVES PROGRAM (WHIP)

Landowners interested in enhancing wildlife habitat can receive assistance through the Wildlife Habitat Incentives Program. The program provides financial incentives to develop habitat for fish and wildlife on private lands. Participants who agree to implement a wildlife habitat development plan receive cost-share assistance from the US Department of Agriculture (USDA) to begin the project. USDA and program participants share the costs. This agreement generally lasts a minimum of 10 years from the date the contract is signed.

For more information, contact your local Natural Resources Conservation Service or Soil and Water Conservation District. Information can also be obtained from NRCS at www.nrcs.usda.gov/OPA/F960PA/whipfact.html

NON-GOVERNMENTAL TECHNICAL ASSISTANCE

Landowners tackling restoration projects typically complain that they need a single point-person, agreement among agencies and a knowledge of the true costs, restrictions and timelines. The landowners who had a navigator through the process tended to have quicker results, less frustration and greater success. In addition to state and federal agencies, non- profits and watershed councils can provide technical assistance and guidance through the maze of regulatory and funding programs. The following groups helped the landowners in this book. Explore your area for other potential sources of assistance.

COOS WATERSHED ASSOCIATION

The Coos Watershed Association is one of the oldest watershed councils in Oregon. A private, not-for-profit corporation, the Association works with willing landowners in Coos County to implement about $750,000 worth of coordinated stream restoration and road rehabilitation projects each year. They have recently begun working with other private groups to help interested landowners sell easements or title for restorable tidal marshes. For more information, contact the Coos Watershed Association at 541.888.5922.

DUCKS UNLIMITED

Ducks Unlimited is a private, nonprofit, international organization dedicated to conserving wetland habitat for waterfowl. It works with landowners and agencies to encourage habitat development and protection on private and public lands, secures funding for habitat development projects and conducts biological research. For more information, contact Ducks Unlimited at 360.885.2011 or www.ducks.org

KLAMATH BASIN ECOSYSTEM FOUNDATION

Klamath Basin Ecosystem Foundation is a new non-profit organization with a mission "to protect, conserve, and restore the natural resources of the Klamath Basin while promoting the long-term sustainability of the region's economy."

For more information contact the Klamath Basin Ecosystem Foundation at 541.850.1717 or wderyckx@cvc.net

NICOLAI-WICKIUP WATERSHED COUNCIL (NWWC)

Nicolai-Wickiup Watershed Council covers east Clatsop County and all the streams and rivers draining into the Columbia River on the Oregon side. Its boundaries span from the edge of Astoria east to the county line. NWWC works to maintain, enhance and restore existing salmon streams and to improve water flow and quality in slough areas diked for flood control (primarily to create agricultural lands and transit routes, e.g. railroad dike). Coordination for the seven Clatsop County watershed councils is handled by the Columbia River Estuary Study Taskforce (CREST). For more information, contact CREST at 503.325.0435 or www.clatsopwatersheds.org/nickolai.htm

OREGON WATER TRUST

Oregon Water Trust is a private, nonprofit group established in 1993, which acquires consumptive water rights from willing water-right holders and converts them to instream water rights. Oregon Water Trust offers water-right holders a variety of incentives: compensation, funding for irrigation efficiency improvements and protection of water rights from non-use. For more information, contact Oregon Water Trust at 503.227.4419 or www.owt.org

SOUTH COAST LAND CONSERVANCY

The South Coast Land Conservancy is a private organization that advises and financially assists South Coast landowners wishing to protect natural characteristics on privately-owned land. The Conservancy helps landowners identify programs or strategies best suited to a particular site or circumstance, provides a centralized liaison with state and federal funders, serves as technical advisor to the landowner for restoration on protected land and occasionally provides funds. For more information, contact the South Coast Land Conservancy at 541.266.7202.

SOUTH COAST WATERSHED COUNCIL

The South Coast Watershed Council is the "umbrella" council for all local watershed councils in Curry County, in Southwest Oregon. The Council includes representatives from 10 "sub-watersheds" — along with state and federal agency representatives. The South Coast Council sets priorities, identifies project opportunities and works with landowners in a million-acre county with 10 watersheds. For more information, contact the South Coast Watershed Council at 541.247.2755 or curswcd@harborside.com

WALLOWA RESOURCES

Wallowa Resources, created in 1996, is a Wallowa County community-based nonprofit organization. Wallowa Resources works with the community to demonstrate the benefits of inclusive, collaborative decision-making, reinvestment in natural resources and education about land stewardship. For more information contact Wallowa Resources at 541 426-8053 or wallowa@oregonvos.net. Their website is www.wallowaresources.org

WATERSHED COUNCILS

Watershed councils are locally organized, voluntary, non-regulatory groups established to improve the condition of local watersheds. The councils offer local residents the opportunity to independently evaluate watershed conditions and identify opportunities to restore or enhance the conditions. The councils forge partnerships between residents, local, state and federal agency staff and other groups. Through this integration of local efforts, the state's watersheds can be protected and enhanced. Contact the Oregon Watershed Enhancement Board for a full listing and map of Oregon Watershed Councils at 503.986.0178 or www.oweb.state.or.us

THE WETLANDS CONSERVANCY (TWC)

The Wetlands Conservancy, founded in 1981, is the only statewide land trust with the specific goal of preserving wetlands. The non-profit land trust conserves, protects and restores wetlands, other aquatic systems, and related uplands through education, research, acquisition and promotion of private and public stewardship. TWC accomplishes its goals through land acquisition, preservation and the Stewardship Program. For more information, contact The Wetlands Conservancy at 503.691.1394 or www.wetlandsconservancy.com

REGULATORY AGENCIES

Private landowners who take initiative to protect wetlands benefit from knowing the relevant laws and regulations. Each requirement has exceptions and may have additional provisions. As illustrated in the preceding stories, a knowledge of permit requirements and timelines is essential to wetland conservation and restoration success. If you are unsure of what laws and requirements might affect you, ask the Division of State Lands or local government planning offices and conservation districts. Some of the agencies and non-profits listed in the preceding pages may also be able to help.

US ARMY CORPS OF ENGINEERS

Under Section 404 of the Clean Water Act, the US Army Corps of Engineers (Corps) and the US Environmental Protection Agency (EPA) share the authority over "discharges of dredged or fill material" into waters of the United States."

Waters of the United States include rivers, streams, estuaries, ponds, lakes and wetlands. "Discharge of dredged or fill material" means placement or movement of any kind of material into or within a wetland or other waters of the United States.

Landowners are required to obtain a permit for dredge and fill activities in waters of the United States, regardless of the amount of fill used or area affected by the project.

US ENVIRONMENTAL PROTECTION AGENCY

The Environmental Protection Agency is responsible for assuring that the Clean Water Act provisions (including Section 404 regulation dredge and fill activities) are implemented through regulatory and non-regulatory programs. The agency's mandate under the Clean Water Act is to protect, maintain and restore the nation's waters. This is done through: state delegated permit authority; supporting development of water quality protection programs through funding to states and tribes; and through technical and financial assistance to states, tribes, watershed organizations and other stakeholder interests (agriculture, forestry, fisheries, etc.) to improve protection of aquatic ecosystems.

US DEPARTMENT OF AGRICULTURE

The Wetland Conservation or "Swampbuster" provision of the Food, Agriculture, Conservation and Trade Act of 1990 is intended to discourage the conversion of wetlands for agricultural purposes. A landowner who drains, dredges, fills or otherwise alters wetlands after November 28, 1990, to make possible the production of agricultural commodity on a wetland that was converted after December 23, 1985 is ineligible for many of the US Department of Agriculture programs discussed in this book.

NATIONAL MARINE FISHERIES SERVICE AND US FISH AND WILDLIFE SERVICE

Wetlands are critical habitat for many species currently named on the federal and state endangered species lists. The federal Endangered Species Act prohibits any person from "taking" endangered or threatened

animal species. An endangered species is any species of plant or animal in danger of extinction throughout all or a significant portion of its range.

The US Fish and Wildlife Service administers the federal Endangered Species Act for terrestrial habitats and inland waters. The National Marine Fisheries Service administers the act for coastal and marine species, including anadromous fish.

DIVISION OF STATE LANDS

The Oregon Division of State Lands administers the state Removal-Fill Law and implements the 1989 Wetlands Conservation Act.

The Removal-Fill Law (ORS 196.800-990) requires people who plan to remove or fill material in waters of the state to obtain a permit from the Division of State Lands. Placement of water control structures and dike structures, channel or bank alteration, land clearing, construction or roads and buildings, and backfilling or restoring "ditched" channels are examples of activities that may require a Removal-Fill Permit or General Authorization (a faster type of permit). Waters of the State are similar to Waters of the United States, as described above for the US Army Corps of Engineers. Activities in Waters of the US and State may require a permit from both the Division of State Lands and the Army Corps.

DEPARTMENT OF ENVIRONMENTAL QUALITY

Section 401 of the federal Clean Water Act requires the Oregon Department of Environmental Quality to certify that the proposed activity does not endanger the health of Oregon's streams and wetlands and to confirm that the plan meets water quality laws and standards. The Department of Environmental Quality (DEQ) issues a Water Quality Certification for Corps permits, and projects permitted by the Division of State Lands must also meet water quality standards set by DEQ. Applicants may be required to incorporate

water quality protection measures such as sediment protection, storm water runoff treatment and protection of fish and wildlife into their plans.

OREGON WATER RESOURCES DEPARTMENT

Under state law, all water is publicly owned. A permit must be obtained to use water or store water in a reservoir from any source, whether it is underground or from lakes and streams. A permit from the Oregon Water Resources Department is required before using ground or surface water. The Oregon Department of Environmental Quality, the Oregon Department of Fish and Wildlife and the Oregon Department of Agriculture review applications for diversions of surface water, or where groundwater is hydrologically connected to surface water from areas where sensitive, threatened and endangered species are present. These agencies may request alterations to better protect fish and wildlife species and water quality.

PROJECT PROFILES AND
WETLAND RESTORATION TECHNIQUES

Each restoration project has its own character. A project whose primary goal is to attract fish may require different strategies than one seeking to create diverse habitat or to restore the historic path of the river, as the following examples demonstrate.

HALL SLOUGH RESTORATION
DeLorenzo Project

Project Goal: To release land-locked cutthroat trout and allow other salmon species to re-colonize Hall Slough.

Objective: The primary task was to repair an existing dike and create a fish passage over it. The passage was designed to provide fish with access to a large pond and several streams draining into Hall Slough. Historically, the area provided habitat for four species of salmon and searun cutthroat trout.

Project Elements:
♦ The existing dike, which maintains the pond, was repaired and stabilized. This involved removing non-native vegetation, rebuilding eroded sections and leveling the dike. Biodegradable erosion-control matting was installed. Once stabilized the dike was planted with native grasses and shrubs.

♦ Four culverts from tributary streams that fed the pond were removed and the natural streambeds were restored.

♦ A county road culvert, which connected the pond and Hall Slough, was replaced with a larger one to better accommodate fish passage. The culvert size was more-than doubled from 18 inches to 38 inches.

♦ A fish passage with a water control structure was created along and over the dike. Biodegradable erosion-control matting and wetland plants were installed in the new fish passage.

Maintenance and Monitoring
1. Before construction, computer data loggers were installed in the pond to record changes in water temperature.

2. Water quality and depth are continually monitored.

3. Fish surveys using live traps were conducted before, during and after construction to determine native species migration patterns.

4. Students from the local high school conduct ongoing water quality and fish species monitoring and research projects at the site.

The Northwest Ecological Research Institute and the property owner continually maintain the dike and fish passage and monitor water quality and wildlife at the site.

MCCOY CREEK RESTORATION
Tipperman Project

Project Goal: The goals are to restore historic meandering to the straightened channel, improve fish habitat by adding structure and complexity to the channel, and planting native streamside vegetation.

Project History
Prior to starting the restoration work, the stream conditions were surveyed. The creek was in poor condition, a straightened channel with few or no pools, little streamside vegetation, and high water temperature. The channel could provide neither safety nor adequate food for juvenile fish. High temperatures, combined with lack of pools, could be lethal to spawners.

Project Elements
♦ A 1937 aerial photograph and an elevation survey of the meadow identified the former channel.

♦ The ditch was blocked with a long gravel berm, creating a large meander.

♦ Portions of the ditch were left connected to the creek for use as refuge for juvenile fish during times of high water.

♦ Salmon Corps workers planted 10,000 willow, cottonwood, and other native plant shoots.

♦ The areas that bulldozers and trucks had flattened and compacted were churned, raked and replanted.

Results
Shortly after restoring the old channel, the water leaving the meadow was 5° to 6° F colder than the water entering the meadow. McCoy Creek now ran in its old channel. New beaver dams sprang up overnight in some places, creating new pools and raising the water table to nourish riparian plants. The underground flow of water and deep pools created by beaver dams played a big role in the nearly immediate temperature reduction.

The underground flow of water began immediately when the dam across the old ditch was completed. Water from the creek filtered underground through the dam and came out a short distance downstream.

One year later, McCoy Creek showed most of the characteristics of a stream reach in good condition. The channel is narrower and deeper. The water is colder. The riparian vegetation around the beaver dams has thickened. And the marsh area has expanded outward some 50 feet. Elsewhere, willow and cottonwood starts show new growth, and other greenery pokes up in streamside gravel. Winter and spring floods have stayed in the channel and floodplain without "blowing out" any of the restoration work.

SPRAGUE RIVER WETLAND RESTORATION **Hines Project**

Project Goal: The project is designed to restore historic wetlands and improve wildlife diversity.

Project Elements:

♦ Swales were reconfigured to create varied depths of water, which would provide diverse habitat and attract a variety of wildlife. Nest islands were created using the excess material from the swale reconfiguration.

♦ A new berm will be constructed to connect the existing dike and road. The berm will encircle the wetland and enable it to hold water.

♦ Sediment will be removed from the oxbow to reconnect it to the river and create more wetland habitat.

♦ New berms and water control structures will be constructed to create wetlands with nesting islands and food plots.

Sprague River Wetland Restoration Plan

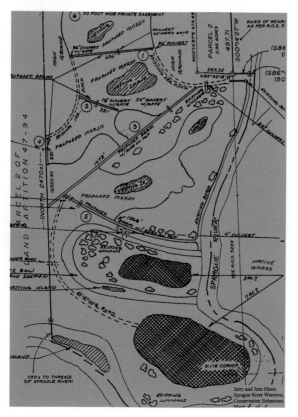

WETLAND RESTORATION TECHNIQUES

Every wetland possesses hydrology, hydric soils and wetland vegetation. Wetland restoration involves returning one or more of these three characteristics to a site. Hydric soils form over a long period of time and the soil characteristics are very difficult to create. For this reason, restorations generally take place where the hydric soils have historically existed, but the hydrology or vegetation has been altered.

Wetland restoration often restores the natural hydrology and topography of a site. Sites where there has been excessive logging, uncontrolled cattle grazing or unrestricted off-road vehicle use can be good candidates for restoration. Projects can include planting, removing cattle, fencing streams and wetlands, removing barriers or restoring the water source and/or other wetland properties.

Nine techniques utilized by the landowners in this book are described in the following pages:

Backfilling Ditches

Ditches are dredged through wetlands to gain wet areas or promote irrigation and move water. The dredged material is often piled along the edge of the ditch in piles, known as spoil banks. These can block sheetflow across the marsh and damage or destroy the wetland.

Backfilling can help restore the natural topography and hydrology, which can lead to at least partial restoration of wetlands impacted by ditch construction.

However, completely removing the material from spoil banks can be difficult. In older ditches, especially those dredged through organic substrates, spoil banks can oxidize, which reduces the volume of backfill. In many cases original elevations cannot be achieved.

Constructing Berms

Constructing small dikes or berms can protect adjacent property from natural flooding of a restored wetland.

Controlling Weeds

Thousands of species of plants have been transported beyond their natural ranges, both intentionally and unintentionally. Many introduced species spread prolifically in environments where predation and competition are limited, pushing out the native flora. The undesired plants or weeds can be pulled manually or mechanically or eradicated with herbicides, grazers or pathogens. They can also wither through manipulation of the hydrology or combination of methods.

Weed control can allow re-establishment of native plant communities. Once the introduced plant populations are well established, removal is a labor intensive ongoing task. A combination of biological, manual and mechanical controls, herbicide application and hydrologic manipulations may be required to eradicate the invasive species. Control using herbicides is not always appropriate. Chemical herbicides may damage the native species as well as hinder the restoration. Hydrological manipulation is not always possible.

Excavating

Excavation can restore natural topography and elevations in order to intercept groundwater, reach an intertidal level or establish wetland hydrology. In some cases, sediment previously deposited in a wetland can be removed to restore the wetland.

When landowners strive to create artificial wetlands, they can displace other habitat (i.e., upland habitat). Excavation and removal of excavated material can be expensive. In some circumstances, it is difficult to predict appropriate excavation depths. Excavation to subsoil leaves poor substrate for plant growth.

Installing Water Control Structures

Restoring natural hydrology is important on restoration projects. A landowner may wish to artificially alter water levels with a water control structure when bullfrogs or other invasive species need to be controlled or when the natural hydrology cannot be re-established. Risers or stop-log structures made of plastic or metal help manage the normal flow of water. An emergency spillway, which is a wide trough-like opening in the side of the dike, should be designed into wetland restoration projects if excess water is expected during flood events. These spillways allow water to pass through without damaging the retention structures in high-water events. Since water management is critical, it is important to determine how much if any water should be controlled in your wetland.

Water control structures generally require a lot of maintenance and monitoring, and may create fish passage barriers and entrapment.

Maintaining the Restored Site

Wetland restoration should be designed to be self-sustaining, requiring little or no maintenance. However, most restored wetlands — particularly those with water-control or earthen structures — require some maintenance. Water-control structures need to be checked regularly. Fallen leaves, twigs or other debris may build up around the mouth of the structure. Debris may obstruct the flow and cause the water level to rise. Inspection of the site, particularly during and after a big storm, will allow the landowner to remove materials before problems develop. Constant control of water levels may be required to insure that fish passage is maintained, and no fish are trapped or stranded.

Ditch plugs, dikes and berms also require some care. Established seedings of grasses should be periodically mowed or burned to prevent woody vegetation from compromising the integrity of the structure. Root growth from woody vegetation allows water to penetrate the earthen structure, contributing to the possibility of a washout.

Reconnecting Floodplains and Restoring Backwaters, Channels and Bends

Restoring a stream to its natural channel and reconnecting the channel and floodplain can reduce sediment load and flooding downstream, raise the water table, lower water temperature and restore fish and wildlife habitat. Productive backwaters, side channels and meanders can serve as a refuge and nursery for young fish and other aquatic life and improve adjacent wetlands. To re-establish these bends in the river, a landowner can modify or remove barriers, such as flood levees, roads, fences, farm tracks and earth banks. Backwaters and side channels that receive water in high flow events help stabilize banks by reducing erosion. They also recharge groundwater, create habitat for a variety of wildlife species and provide refuge for fish during flooding.

A straightened stream can be reconnected to parts of its former meandering channel by removing dikes or levees that keep it in check. The historic channel can be identified from old photos or by the presence of a residual line of vegetation. This may require heavy equipment and assistance from a geomorphologist or hydrologist.

Removing Culverts

Removing or repairing culverts can be an effective way to increase fish habitat. Culverts can block fish passage by constricting flows, collecting debris that plugs passage and forcing the water to find another path, often one that a fish cannot follow.

In many coastal areas, roads have been built across tidal creeks, separating tidal wetlands from the estuary. Frequently, tidal creek flow is maintained by installation of culverts or pipes, that pass beneath the road. However, these pipes are sometimes too small to allow full tidal flushing of wetlands. Subsequently, the area of tidal wetland that was flooded by the tides prior to culvert installation is reduced. Furthermore, if a culvert is installed with the bottom of the culvert above the level of the creek bed, the culvert will act as a weir, holding water on the wetland. This may cause loss of plants and, in some settings hypersalinity.

Removing Tile

Tile breaking involves removing a section of underground agricultural tile that is draining a wetland basin. Drain tile or field tile is usually made of clay or perforated plastic and buried at a depth of two to six feet. Generally, a backhoe is used to remove or crush a 25- to 50-foot section of tile downstream of the basin. The downstream end or outlet pipe is then plugged with a bag of redi-mix concrete or clean clay fill and the trench is filled. Sometimes, a portion of unperforated tile or riser is connected to the downstream end of the tile line and brought to the surface in order to control the water level. Water will fill the wetland basin until it reaches the mouth of this riser, where it will then flow back through the tile line into the ditch.

LANDOWNER
HURDLES AND DIFFICULTIES

While each restoration story in this book has its own set of circumstances, specific to a piece of land and region of the state, many of the landowners share common frustrations. Here are the six most common hurdles and difficulties described by landowners:

1. The process needs to be more user-friendly.

◆ Agency officials and programs use too much jargon.

◆ Information and advice is often impractical and does not work within a working farm or ranch.

◆ Project designs are too restrictive and do not allow for creativity in design and implementation.

◆ Programs focus primarily on fish and endangered species without necessarily promoting habitat restoration and non-game species.

◆ The number and length of meetings can seem endless and pointless, especially when landowners are asked to travel long distances to attend.

2. The process is unclear.

◆ There is no road map or outline of the process from start to finish.

◆ Landowners sometimes find themselves mediating between federal and state resource agencies over conflicting programs and philosophies.

◆ Landowners are often confused about hierarchies among agencies, permits and laws.

◆ There is no central clearinghouse for a landowner to call with questions, clarifications or status of permits.

◆ Complying with programs requires a host of permits and clearances with agencies who use different definitions for work that will result in similar outcomes.

3. The timeline is uncertain.

◆ Timeline projections are unrealistic. The entire process can take up to three times longer than anticipated to complete.

◆ Appraisal and survey delays can put landowners into tight financial binds.

4. Agencies are understaffed.

◆ Taking a project from start to finish requires a lot of guidance, technical assistance and handholding by someone familiar with the process, wetland restoration, fish and wildlife, regulations and permits. Most agencies do not have enough staff to provide the assistance required by landowners.

◆ Some agency staff have a very good knowledge of fish and wildlife needs or restoration, but very little understanding of farming or hands-on land management experience. This can lead to miscommunication and misunderstandings.

◆ Project success can be dependent on personalities and staff expertise and interest.

5. There needs to be more flexibility in management choices and restrictions.

Some landowners find the bureaucratic maze restricts their ability to mow, burn or actively manage their land.

6. There needs to be more flexibility and options in compensations and easements.

- Federal easements do not allow landowners to take deductions for added values, such as timber.

- Under the Wetlands Reserve Program landowners cannot negotiate for the type of easement. The current form leaves the landowner with no recorded easement.

- Programs can penalize farmers for conversion of farmland to wetland. Once land in Oregon enters the Wetlands Reserve Program (WRP), it may lose its "farm use" tax status. Landowners must then pay taxes on fair market value, unless the Oregon Department of Fish and Wildlife signs off on a wildlife management plan.

- US Department of Agriculture administers two different payment programs: the Farm Service Agency's (FSA) "Debt for Nature Program" and the Natural Resources Conservation Service's Wetlands Reserve Program (WRP) have different compensation, standards and requirements. The Debt for Nature Program is only available to ranchers who have defaulted on their loan. It does not provide technical assistance or funding for restoration. The WRP is open to all and provides a lower rate of compensation. WRP provides both technical assistance and funding for restoration.

RECOMMENDATIONS ON HOW TO IMPROVE PROGRAMS

Landowners and natural resource agency staff suggest the following nine improvements to incentive and technical assistance programs:

◆ Streamline the regulatory and permit processes to be more friendly to the partners. These are not publicly owned lands and even though it is government money, the regulatory process for private partners should be less complicated and intimidating.

◆ Staff from all federal and state natural resource agencies need to work out their differences and set common landscape, fish and wildlife, and restoration goals before introducing private landowners to the process.

◆ Allocate money to local offices and eliminate the regional ranking process. Too many hands in the pie dough never gets the pie made.

◆ Retain agency participation in management after project completion. Many of these projects are artificially maintained wetlands and will require management activity from time to time.

◆ Develop a good understanding of the physical, biological and social landscape of an area before beginning to work on projects with landowners.

◆ Check with landowners after project completion to see what worked, what didn't and what, if any, retrofits are needed.

◆ Improve communication. All parties need to commit to the annual meeting to review management goals and activities required by the programs.

◆ Consolidate all federal programs involving compensation into a single office.

◆ Standardize compensation, standards and requirements between all USDA payment programs.

LANDOWNER ADVICE TO OTHER LANDOWNERS

Many of the landowners interviewed experienced delays and frustrations. However, some view themselves as pioneers, blazing a path for others interested in working with government programs. They provide the following advice to fellow landowners:

◆ Find a non-agency advocate familiar with the process to assist you.

◆ Assume it will take a lot longer than you think to complete the project.

◆ Check with all the different agencies before choosing one design. This may save a lot of time, money and frustration in the end.

◆ Explain the project to neighbors and members of the community. If possible, engage them in the project.

◆ Subscribe to the *Capitol Press* to learn more about incentive and grant programs.

◆ Take good photos before, during and after you begin work.

◆ Learn all the restoration vocabulary and jargon before you start the project. It will save you time in the long run.

◆ Develop thick skin. The disapproval of neighbors or other community members, who do not understand or support your project, can cause you discomfort.

The production of this book was accomplished with the talent and assistance of the following people:

Writing:	Esther Lev
Editing:	Dawn Robbins
Design and layout:	Laurie Causgrove
Illustration:	Steve Katagiri
Photography:	Madeleine Blake
	Steve Roundy
	Deb Stoner
	Richard Wilhelm
Additional photos provided by:	Curt Mullis
	Teresa DeLorenzo
Printing:	Publisher's Press, Salt Lake City, Utah
Production Assistance:	Bob Smith, BookPrinters Network

Photograph credits:

Madeleine Blake – pages 25, 28, 31, 32

Teresa DeLorenzo – pages 6, 7, 66

Curt Mullis – pages 26, 27, 30, 34, 35, 70

Steve Roundy – pages 40, 42, 43, 44, 46, 47

Deb Stoner – page 4

Richard Wilhelm – pages xii, 2, 3, 8, 10, 11, 12, 14, 15, 16, 18, 19, 20, 22, 23, 36, 38, 39, 50, 54, 55, 56, 63, 65, 72